T0234468

Bionic Gliding Underwater Robots

Underwater robots play a significant role in ocean exploration. This book provides full coverage of the theoretical and practical aspects of bionic gliding underwater robots, including system design, modeling control, and motion planning.

To overcome the inherent shortcomings of traditional underwater robots that can simultaneously lack maneuverability and endurance, a new type of robot, the bionic gliding underwater robot, has attracted much attention from scientists and engineers. On the one hand, by imitating the appearance and swimming mechanisms of natural creatures, bionic gliding underwater robots achieve high maneuverability, swimming efficiency, and strong concealment. On the other hand, borrowing from the buoyancy adjustment systems of underwater gliders, bionic gliding underwater robots can obtain strong endurance, which is significant in practical applications. Taking gliding robotic dolphin and fish as examples, the designed prototypes and proposed methods are discussed, offering valuable insights into the development of next-generation underwater robots that are well suited for various oceanic applications.

This book will be of great interest to students and professionals alike in the field of robotics or intelligent control. It will also be a great reference for engineers or technicians who deal with the development of underwater robots.

Junzhi Yu is a professor in the Department of Advanced Manufacturing and Robotics, Peking University, China, and a guest researcher at the Institute of Automation, Chinese Academy of Sciences. His research interests include bionic robots, intelligent control, and intelligent mechatronic systems. He has authored or co-authored five monographs and published more than 100 science citation index papers in prestigious robotics and automation–related journals. He has successively been listed among the Most Cited Researchers in China between 2014 and 2020. He is a fellow of the Institute of Electrical and Electronics Engineers.

Zhengxing Wu is currently a professor with the State Key Laboratory of Management and Control for Complex Systems at the Institute of Automation, Chinese Academy of Sciences. His current research interests include bio-inspired robots and intelligent control systems.

Jian Wang is currently an assistant professor with the State Key Laboratory of Management and Control for Complex Systems at the Institute of Automation, Chinese Academy of Sciences. His research interests include bio-inspired underwater robots and intelligent control systems.

Min Tan is currently a professor with the State Key Laboratory of Management and Control for Complex Systems at the Institute of Automation, Chinese Academy of Sciences. He has published more than 200 papers in journals, books, and conference proceedings. His research interests include robotics and intelligent control systems.

Bionic Gliding Underwater Robots

Design, Control, and Implementation

Junzhi Yu, Zhengxing Wu,
Jian Wang, and Min Tan

CRC Press
Taylor & Francis Group
Boca Raton London New York

CRC Press is an imprint of the
Taylor & Francis Group, an **informa** business

This book is published with financial support from the National Natural Science Foundation of China under Grant 61725305, Grant 62033013, Grant U1909206, and Grant T2121002.

First edition published 2023
by CRC Press
6000 Broken Sound Parkway NW, Suite 300, Boca Raton, FL 33487-2742

and by CRC Press
4 Park Square, Milton Park, Abingdon, Oxon, OX14 4RN

CRC Press is an imprint of Taylor & Francis Group, LLC

© 2023 Junzhi Yu, Zhengxing Wu, Jian Wang, and Min Tan

Library of Congress Cataloging-in-Publication Data
Names: Yu, Junzhi (Writer on robotics), author. | Wu, Zhengxing, 1989- author. |
    Wang, Jian, 1993- author. | Tan, Min, 1962- author.
Title: Bionic gliding underwater robots : design, control and implementation /
    Junzhi Yu, Zhengxing Wu, Jian Wang, Min Tan.
Description: First edition. | Boca Raton, FL : CRC Press, 2023. |
    Includes bibliographical references.
Identifiers: LCCN 2022027761 (print) | LCCN 2022027762 (ebook) |
    ISBN 9781032389134 (hbk) | ISBN 9781032389141 (pbk) | ISBN 9781003347439 (ebk)
Subjects: LCSH: Robotic fish. | Hydrofoils. | Robots—Control systems.
Classification: LCC TJ211.34 .Y83 2023 (print) | LCC TJ211.34 (ebook) |
    DDC 629.8/92—dc23/eng/20220902
LC record available at https://lccn.loc.gov/2022027761
LC ebook record available at https://lccn.loc.gov/2022027762

ISBN: 978-1-032-38913-4 (hbk)
ISBN: 978-1-032-38914-1 (pbk)
ISBN: 978-1-003-34743-9 (ebk)

DOI: 10.1201/9781003347439

Typeset in Minion
by Apex CoVantage, LLC

Contents

CHAPTER 1

Development and Control of Underwater Gliding Robots

A Review

1.1 INTRODUCTION

The ocean occupies 71% of the earth's surface area and contains abundant biological resources, mineral resources, and vast space resources. Therefore, with the vigorous development of economy and technology, ocean exploration has received widespread attention from all over the world [1], [2]. However, it is very challenging to explore the ocean due to the complex environment. The reasons include many aspects. First, the ocean area is too large for humans or underwater vehicles to traverse all areas owing to limited energy. Second, accompanied by unpredictable ocean currents and extreme weather, the underwater environment is intricate, which demands great stability and maneuverability for underwater exploration. Third, the problems of underwater communication and signal transmission have not been effectively solved so far. The current devices, such as the Global Positioning System (GPS), can only be transmitted in the air, and the underwater acoustic communication distance is not long enough. In summary, ocean exploration requires long endurance, high maneuverability, strong stability, and enough load capacity.

DOI: 10.1201/9781003347439-1

Underwater unmanned vehicles (UUVs) are mainly used today to conduct ocean exploration, including autonomous underwater vehicles (AUVs) and remote-operated vehicles (ROVs) [3]–[5]. The main difference is that ROV needs cable operation while AUV is a completely autonomous type. Therefore, an AUV is more suitable for wide-area ocean observation. As an AUV with long endurance characteristics, the underwater glider (UG) is the most effective underwater unmanned platform for ocean observation in recent years [6]. Compared with other underwater vehicles, an UG can achieve gliding motion by changing the net buoyancy. It has the advantages of low energy consumption, a wide range of navigation, long endurance, and low noise. As a new type of marine monitoring and observation platform, the UG can be equipped with a series of research instruments to measure and transmit environmental data, such as depth, salinity, temperature, and ocean current in real time. It plays an important role in marine environmental monitoring, resource detection, and disaster forecasting.

Since 1989, many scientists have carried out a lot of research in the field of underwater gliding robots. With the development of economy and technology, a large number of new underwater gliding robots (UGRs) have emerged in recent years, and they have made certain innovations and improvements in terms of appearance, driving mechanism, and energy power system. Although there is much literature investigating UGs [7]–[11], most of them compared the performances of traditional UGs. On one hand, there were few overviews for new types of UGRs. On the other hand, the most recent studies were published 4 years ago. [7], [8] introduced the UG developments by comparing the representative UGs, that is, Slocum, Spray, and Seaglider, with each other and discussed the role of gliders in ocean research infrastructure. From the point of working depth and payload capacity, various UG platforms were reviewed by Javaid et al. [9]. Besides, control strategies of UGs with emphasis on their actuation system were discussed by Ullah et al. [10]. In addition, a comprehensive review of UG was presented by Shen et al. [11], the emphases of which lie on the research progress of the single UG platforms and the corresponding core techniques. Although some progress has been made in the UG field, there was limited review summarizing the aspects of the variety of UGRs.

From a novel perspective, this chapter focuses on UGRs, including traditional UGs, hybrid-driven UGs, thermal UGs, bio-inspired UGs, and others. The overview framework is shown in Figure 1.1. This chapter aims

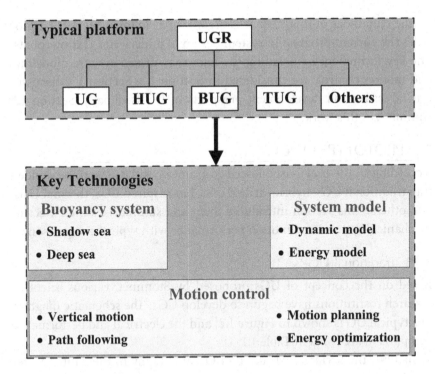

FIGURE 1.1   The overview framework of underwater gliding robots.

to provide researchers with a comprehensive overview of UG research and points out the future development prospects of UGRs. To the best of our knowledge, few studies focused on the classification of UGRs, especially on bio-inspired UGs, which are crucial for practical applications of new-type UGRs. The main contributions of this chapter are summarized as follows:

- This chapter comprehensively summarizes the current platform research status of UGRs, including traditional UGs, hybrid-driven UGs, bio-inspired UGs, thermal UGs, and other types of UGs.

- As key technologies of UGRs, we outline the research status of buoyancy-driven systems and control methods of UGRs, including system modeling, motion control, and gliding optimization.

- We analyze the current research bottlenecks and future development of UGRs, providing certain guidance and deployment suggestions for researchers in future research.

The remainder of this chapter is organized as follows: Section 1.2 provides the current progress in system design of underwater gliding robots. The key technologies, including the buoyancy-driven system, modeling, and motion control, are considered in Section 1.3. Section 1.4 discusses critical issues and development prospects of UGRs. Finally, Section 1.5 summarizes this work.

1.2 PROTOTYPE OF UGRs

According to the previously described framework, UGRs can be divided into traditional UGs, hybrid-driven UGs, bio-inspired UGs, thermal UGs, and others. This section introduces them and summarizes their electromechanical systems and motion performance with typical comparisons.

1.2.1 Traditional UGRs

Based on the concept of UGs proposed by Stommel, various scientific research institutions have begun to develop UGs. The schematic diagram of a typical UG is shown in Figure 1.2, and the electrical and performance comparison are listed in Table 1.1.

In 1991, the Scripps Institution of Oceanography in the United States developed an autonomous Lagrangian circulation explorer (ALACE) [12] as a profile buoy, which is a length of about 1.07 m, an antenna length of 0.7 m, a shell made of T6061 aluminum alloy, and a weight of about 23 kg. Besides, the testing maximum water depth is 2000 m. Through satellite positioning, ALACE can complete long-term fixed-point observation and communication. This UG has carried out a large number of tests to maximize energy efficiency and developed a relatively complete buoyancy-driven system, which has laid an important foundation for the subsequent development of the UG.

In 1992, Tokai University designed an UG named ALBAC [13], whose length and weight are about 1.4 m long and 45 kg, respectively. Via horizontal and longitudinal movements of a slider inside, the pitch and roll motions can be achieved. It should be noted that the glider starts the surfacing motion by throwing a heavy mass, which is about 3L, so the glider can only finish a single round of gliding motion after one launch. Moreover, ALBAC completed the test verification in Ashinoko Lake and Suruga Bay and explored the control effect of the center of gravity slider on the glide angle, which can range from 15° to 30°.

Since 2001, the University of Washington has developed a series of underwater gliders named Seaglider [16]. This type of glider adopted the

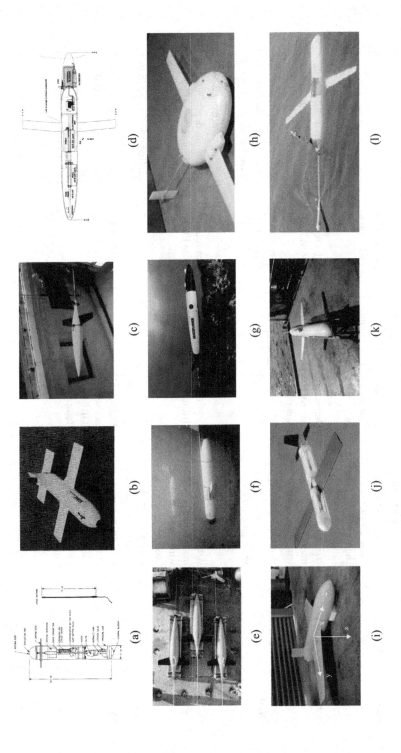

FIGURE 1.2   The overview of the traditional UGs. (a) ALACE [see [12]]. (b) ALBAC [see [13]]. (c) Deepglider [see [14]]. (d) Spray [see [15]]. (e) Seaglider [see [16]]. (f) Slocum [see [17]]. (g) Sea-explorer [see [18]]. (h) ROGUE [see [19]]. (i) UG of Kyushu University [see [20]]. (j) ALEX [see [21]]. (k) Sea-wing1000 [see [22]]. (l) Sea-wing7000 [see [23]].

TABLE 1.1 Traditional UGs

Year	Glider	Mechanical characteristics	Electrical system	Energy	Motion performance
1991	ALACE [12]	size: φ0.17×1.07 m mass: 23 kg shell: T6061 Aluminum alloy	Depth sensor, GPS, communication module	Alkaline battery, 22 Ah and 1000 kJ	Maximum depth: 2650 m
1992	ALBAC [13]	size: φ0.24×1.4 m mass: 45 kg wingspan: 1.0 m	Depth sensor, velocity sensor, ranging sensor, magnetic sensor	Ni-Zn battery, 100 w, 13.6 V, 7 Ah	Maximum depth: 300 m Steady gliding angle: 18.7° Steady pitch angle: 13°
Since 2001	Seaglider [16]	size: φ0.3×1.8 m mass: 52 kg wingspan: 1.0 m	Depth sensor, GPS, CTD, ADCP, IMU	Lithium primaries, 24 V and 10 V packs, 10-MJ	Maximum depth: 1000 m Maximum endurance: 4600 km Steady gliding angle:16–45° Turning rate: 0.2–0.6°/s Gliding velocity: 0.2–0.25 m/s
2007	Deepglider [14]	size: φ0.3×1.8 m mass: 52 kg wingspan: 1.0 m volume change: 1.32 L shell: Carbon fiber composite material	Depth sensor, GPS, Iridium	Lithium sulfuryl chloride Energy 17.0 MJ	Maximum depth: 6000 m Maximum range: 8500 km Maximum endurance: 380 days Steady gliding angle:17° Gliding velocity: 0.225 m/s
2001	Spray [15], [24]	size: φ0.2×2 m mass: 50.2 kg wingspan: 1.0 m shell: T6061 aluminum alloy	Altimeter, pressure sensor, GPS, ORBCOMM, temperature sensor	Lithium battery, 12 kg, 13 KJ	Maximum depth: 1500 m Maximum range: 6000 km Steady gliding angle:18–36° Gliding velocity: 0.2–0.4 m/s

Year	Name	Size/Mass	Sensors	Battery	Performance
Since 2001	Slocum battery [17], [25], [26]	size: φ0.21×1.5–2 m; mass: 52 kg; wingspan: 1.2 m with 45° sweep angle	Altimeter, pressure sensor, GPS, IMU, ORBCOMM, temperature sensor	Alkaline battery, 8 MJ, 18 kg	Maximum depth: 30, 50, 100, 200, 350, 1000 m; Maximum range: 2300 km; Gliding velocity: 0.25–0.4 m/s
2014	Seaexplorer [27], [28]	size: φ0.25×2 m; mass: 59 kg; volume change: 0.5 L	Altimeter, CTD, IRIDIUM Modem	Lithium Rechargeable/primary batteries	Maximum depth: 1000 m; Maximum range: 1300/3200 km; Maximum endurance: 64/160 days
2007	UG of Kyushu University [20]	size: φ0.2×1.2 m; mass: 42.8 kg; wingspan: 1.2 m with 28° sweep angle; volume change: 1.4 L	Depth sensor, accelerometer	-	Steady gliding angle:33.4°; Gliding velocity: 0.67 m/s
2008	ALEX [21]	size: 0.83×0.17 m; mass: 4.35 kg; wingspan: 0.83 m; volume change: 0.5 L	Depth sensor, GPS, magnetic compass	Ni-Cd batteries	Maximum depth: 5 m; Roll rate: 25°/s; Gliding velocity: 0.2–1 m/s
2011	Sea-wing 1000 [22]	size: φ0.22×2 m; mass: 65 kg; wingspan: 1.2 m	Iridium satellite, RF, GPS, altimeter, sonar, CTD, compass, oxygen sensor	Lithium primary/secondary batteries	Maximum depth: 1200 m; Maximum range: 500 km; Turning rate: 0.6°/s; Vertical velocity: 0.05–0.25 m/s
2017	Sea-wing 7000 [23]	size: φ0.3×2.57 m; mass: 140 kg; volume change: 4.3 L; shell: Carbon fiber composite material	Iridium satellite, RF, GPS, altimeter, CTD, compass	Lithium primary batteries, 13.6 MJ, 14.9 kg	Maximum depth: 7000 m; Maximum range: 2000 km; Turning rate: 0.6°/s; Maximum Vertical velocity: 0.6 m/s

streamline shape with a low resistance. A typical one was about 1.8 m long with a weight of 52 kg and had a variable buoyancy volume of 850 ml. By improving the buoyancy drive system, it had stable long-term operation ability and successfully completed the test verification in the Puget Sound. In addition, by carrying a hydrophone, its improved version of Seaglider™ realized near-real-time acoustic monitoring of beaked whales and other cetaceans [29]. In 2007, the research institution increased the pressure resistance of Seaglider by using a composite pressure hull of thermoset resin and carbon fiber and developed a deepsea glider deepglider [14], which can dive to a depth of 6000 m.

In 2001, the Scripps Institution of Oceanography in the United States developed an UG named Spray [15] with a length of about 2 m, a wingspan of 1.2 m, and a weight of about 50 kg. It was equipped with a buoyancy-driven system with a maximum buoyancy change of 900 ml at the rear of the body. Using the ORBCOMM communication module, two-way communication can be realized at 25 bytes per minute. The Spray successfully observed internal waves in the Monterey underwater canyon using onboard sensors and tides. Since 2004, Spray UGs have been deployed for more than 28,000 days, traveling over 560,000 km and delivering over 190,000 profiles [24].

Since 2001, Teledyne Webb Research of the United States has begun to develop the Slocum series of UGs [7], [8], [17], [30], mainly including the Slocum battery and Slocum thermal. The former mainly relied on electricity to drive the buoyancy adjustment system, and the latter used the temperature difference to change the buoyancy. Furthermore, the Slocum battery could be divided into two types of shadow and deep sea, which can dive to depths of 30, 50, 100, 200, 350, and 1000 m [26]. Correspondingly, the different buoyancy system was designed to meet the requirements of different pressure environments. In addition, Alvarez *et al.* conducted some research on the annular wing shape of Slocum [31]. The Slocum UG is one of the most mainstream application-type UGs in the world and has completed a variety of marine operations.

The scientists from France designed an UG, Sea-explorer [18], [27], which was about 2 m long and adopted the wingless design concept. By using short fins instead of long wings, the wing break during working in seaweed plastic debris or fishing nets was avoided effectively. By carrying loads such as acoustic doppler current profiler, a profile observation was achieved in the northwestern Mediterranean Sea, and the effectiveness of data collected by the glider was verified by comparing them with the ship [28].

In 2000, Princeton University in the United States developed a small laboratory-level glider [19]. It adopted a flat shape design with a large wingspan. It was about 18 cm long and had a single wingspan of 2 cm. It is buoyantly driven by suction and drainage. Based on this model, more detailed gliding motion modeling and glide angle control are carried out.

The Kyushu University in Japan designed an UG [20] similar to the shape of an aircraft. It was about 1.2 m in length. A suction-and-drainage and a movable slider mechanism were used to realize buoyancy drive and change the center of gravity. However, the buoyancy drive has the problem of water leakage. Therefore, it can only glide to 3-m-depth underwater. In 2008, an underwater glider ALEX [21] with a length of about 0.83 m was presented by the Osaka Prefecture University in Japan. The glider had independently controllable main wings for the use of oceanographic survey and worldwide-scale monitoring of marine environment. Via applying the computational fluid dynamics (CFD) technology and experiments, and the results showed that the performances of the controllable airfoil were better than that of the fixed airfoil.

Since the beginning of the 20th century, the Shenyang Institute of Automation, Chinese Academy of Sciences has carried out a lot of research on UGs. Initially, for shallow-water gliders [32], the suction and drainage mechanism was used to adjust the net buoyancy. Then, from 2013, it was devoted to deepsea gliders and has conducted a lot of technical research. The typical representatives were the Sea-wing 1000 [22] and the Sea-wing 7000 [23], of which the Sea-wing 7000 used an innovative composite carbon fiber material, which was optimized in terms of machinery, hardware, and software from the energy perspective. The maximum diving speed reached was 0.6 m/s. During the upward gliding process, as long as the absolute value of the speed depth was less than 0.3 m/s, the glider would drain the oil, so the glider could maintain a stable speed. This strategy can effectively reduce system power. In February 2017, the Sea-wing 7000 successfully completed the task of diving to a 6329 m depth in the Mariana Trench, which is a world record.

1.2.2 Hybrid-Driven UGRs

With the aid of a propeller on AUVs, the hybrid-driven underwater glider (HUG) can obtain both high speed and strong endurance. The typical HUGs are figured in Figure 1.3, and their detail are listed in Table 1.2.

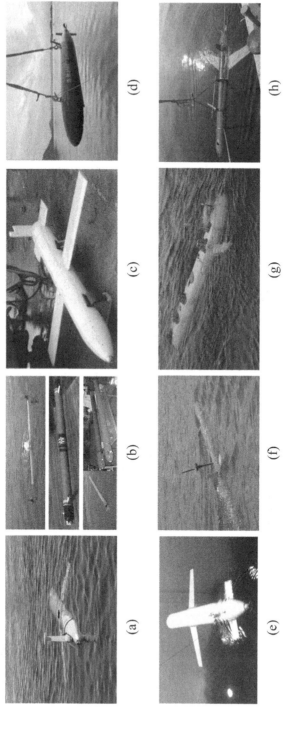

FIGURE 1.3 The overview of the HUGs: (a) Slocum G3 [see [33]]. (b) FOLAGA [see [34]]. (c) Sterne [see [35]]. (d) Sea-whale 2000 [see [36]]. (e) Sea-wing-H [see [37]]. (f) Petrel [see [38]]. (g) Petrel-X [see [39]]. (h) Smartfloat [see [40]].

TABLE 1.2 Hybrid-Driven UGs

Year	Glider	Mechanical characteristics	Electrical system	Energy	Motion performance
2011	Slocum [26]	size: φ0.22×1.5 m mass: 56.3 kg wingspan: 1.2 m	Pressure sensor, Altimeter, attitude sensor, GPS, RF	Alkaline battery, 8 MJ, 18 kg	Maximum depth: 200, 1000 m Maximum range: 1200/3000/13000 km Gliding velocity: 0.5 m/s Propulsion velocity: 1 m/s
2006	FOLAGA III [34]	size: φ0.14×2 m mass: 30 kg volume change: 0.5 L	Pressure sensor, attitude sensor, GPS, acoustic Micro-modem	lead acid battery 12 V, 72 Ah	Maximum depth: 100 m Maximum endurance: 8 h Jet velocity: 1.01 m/s Propulsion velocity: 2.02 m/s
2001	Sterne [44]	size: φ0.6×4.5 m mass: 990 kg volume change: 4 L	–	–	Maximum range: 120 mile Gliding velocity: 1.3 m/s Propulsion velocity: 1.8 m/s
2019	Sea-whale 2000 [36], [45]	size: φ0.35×3 m mass: 215 kg volume change: 3 L	Pressure sensor, attitude sensor, DVL, GPS, RF, Iridium satellite	Lithium primary batteries, 20 KWh	Maximum depth: 2000 m Maximum range: 1500 km Gliding velocity: 0.5 m/s Propulsion velocity: 1.2 m/s
2016	Sea-wing- H [37]	size: φ0.22×2.2 m mass: 67 kg volume change: 1 L wingspan: 1.2 m	Pressure sensor, attitude sensor, Altimeter, DVL, GPS, RF, Iridium satellite	Lithium primary batteries, 13.8 kg	Maximum depth: 300 m Maximum range: 1500 km Gliding velocity: 0.5 m/s Propulsion velocity: 1 m/s
2013	Petrel-II [38]	size: φ0.22×2.3 m mass: 70 kg volume change: 1.4 L wingspan: 1.2 m	Pressure sensor, attitude sensor, Altimeter, CTD, GPS, RF, Iridium satellite	Lithium primary batteries, 18.9 MJ	Maximum depth: 1500 m Maximum range: 1500 km Gliding velocity: 0.5 m/s Propulsion velocity: 1.6 m/s

(Continued)

TABLE 1.2 (*Continued*) Hybrid-Driven UGs

Year	Glider	Mechanical characteristics	Electrical system	Energy	Motion performance
2019	Petrel-X [39]	size: φ0.63×4.07 m mass: 398.55 kg volume change: 10 L wingspan: 2.34 m	Pressure sensor, attitude sensor, Altimeter, CTD, GPS, RF, Iridium satellite	Lithium primary batteries, 18.1 MJ	Maximum depth: 10000 m Maximum range: 1000 km Gliding velocity: 0.86 m/s Propulsion velocity: 1.1 m/s
2019	Smartfloat [40]	size: φ0.3×3.55 m mass: 170.6 kg volume change: 0.22 L shell: Carbon fiber composite material	Pressure sensor, attitude sensor, Altimeter, GPS, Iridium satellite	Li/SOCl2, 35 kg, 70 MJ	Maximum depth: 4000 m Gliding velocity: 0.3–0.5 m/s
2019	Nezha III [50]	size: φ0.16×0.58 m mass: 18 kg volume change: 0.22 L wingspan: 1.65 m	Pressure sensor, attitude sensor, RF, GPS	LiPo batteries, 22 V, 24 AHJ	Maximum depth: 50 m Maximum vertical velocity: 0.94 m/s

In recent years, Slocum has mainly launched a series of hybrid-driven UGs [7], [8], [33]. By using different battery types, such as alkaline batteries, rechargeable batteries, and lithium batteries, Slocum G3 can complete the range of 350–1200 km/700–3000 km/3000–13,000 km [33]. With the aid of GPS, pressure sensors, altimeters, and other sensors, it can complete various underwater tasks. Besides, by using RF, Iridium (RUDICS), ARGOS, Acoustic Modem, it owns the communication ability. Regarding the motion performance, a gliding speed of 0.35–0.5 m/s and a propeller propulsion speed of 1 m/s can be achieved.

Between 2004 and 2006, Italy and Spain developed three types of underwater vehicles, named FOLAGA [34], [41]–[43], of which the first and second generations both applied a propulsion jet pump. In detail, two jet propulsion were equipped at the head and tail cabin to control the steering, and the middle jet thruster is used for vertical movement. Furthermore, the third generation is a hybrid-driven UG, which is equipped with a propulsion device powered by a propeller and a jet pump. The gliding motion was realized by suction and drainage and a worm gear. It can achieve the speed of 1 m/s and 2 m/s with the jet pump and propeller, respectively. At the same time, the internal cabin contained a mobile battery pack that can change the body attitude. In particular, although the energy efficiency of the jet pump seemed lower than that of the traditional propeller, it had the advantages of low realization cost and simple structure and would not cause damage to the surrounding creatures.

In 2001, French ENSTA (École nationale supérieure d'ingénieurs) Bretagne designed a hybrid-driven UG, Sterne [44], which was about 4.5 m long with a mobile battery pack inside to adjust the pitch attitude. Based on the prototype, the robust autonomous navigation method was studied [35]. In addition, the research institution also developed a proportionally reduced model with a length of about 1.8 m to conduct verification. Compared with a same-sized UG, it had twice the amount of buoyancy adjustment.

The Shenyang Institute of Automation, Chinese Academy of Sciences has developed two hybrid-driven UGs, Sea-Whale 2000 [36], [45] and Sea-wing-H [37]. Concretely, Sea-Whale 2000 was designed for meeting the long-term deepsea mobile survey needs in the South China Sea. It was about 3 m long, weighed about 215 kg, and adopted a main wingless design. Regarding Sea-wing-H, it adopted a folding propeller design with a length of about 2.2 m [46], [47]. Compared with the traditional fixed propeller, it was verified that the design can significantly reduce the sailing resistance and improve sailing efficiency.

In 2007, Tianjin University designed the first-generation hybrid-driven UG Petrel [48] and in 2013 developed the second-generation Petrel-II with a length of 2.3 m long [38], [49]. Based on this prototype, the stability of the hybrid drive was proved, and the sea trial in the South China Sea was successfully conducted. In addition, in April 2018, the developed Petrel-10000 dived to 8213 m for the first time, breaking the deepsea world record held by Sea-wing [11]. Furthermore, in 2019, based on the dual–buoyancy adjustment system, Tianjin University developed a deepsea glider, Petrel-X, with a length of about 4.07 m, and dynamic modeling and motion analysis were also investigated and analyzed [39].

By combining the motion characteristics of ARGO profile buoy and UG, a multimodal hybrid-UG was developed by Shanghai Jiaotong University and realized a multimodal drift mode and gliding mode by changing the internal movable mass. When the scientists instruct the vehicle to cross the eddy currents, fronts, filaments, or some stirring regions, the HUG could exploit a float motion to drift with the currents, making measurements and transforming to operate as an UG [40], [51]. In addition, the research institution further developed a hybrid-driven UG NEZHA [52] by borrowing from the characteristics of fixed-wing UAVs, rotary-wing UAVs, and UGs. Based on this, a water–air integration glider NEZHA III was designed. This glider adopted folding rotors. On one hand, the rotors can be stowed underwater and opened when it was about to surface. On the other hand, they can also be opened when it needs to vertically ascend and descend. Moreover, the effectiveness of the designed platform was successfully validated via the lake test of gliding and aerial flight [50]. In 2021, Shanghai Jiao Tong University developed the newest hybrid UG that integrates profile buoys, UGs, and a propeller-driven mode. The glider can realize Argo mode, zigzag gliding mode, and propeller-driven mode, which can be applied for multiple water-column sampling [53].

1.2.3 Bio-Inspired UGRs

Bio-inspired aquatic robots have attracted more interest from scientists in recent years and have played more important roles in applications of underwater exploration, observation, and operation [54]–[56]. Although the robotic fish owns excellent motion performance [57], [58], it is difficult to achieve long-range motion, which are limited by battery capacity, which has become an important factor hindering their application in ocean exploration. On the contrary, the UG has strong endurance characteristics, but its maneuverability is poor. Therefore, many researchers began to introduce the driving mechanism of UGs into bionic underwater robots, that is, bionic UGs. The state-of-art prototypes are shown in Table 1.3.

Nanyang Technological Institute of Singapore has carried out research on manta-like underwater gliders. Through the development of three generations of principle prototypes, RoMan I–III, the high-mobility flapping

TABLE 1.3 The Bio-Inspired UGs

Item	Biological creatures	Mechanical characteristics	Max gliding velocity	Prototype
SIA-BUG [59]	snake	size: φ0.106×1.6 m mass: 11.3 kg shell: T7050 aluminum alloy	–	
NTU-RoMan II [60]	manta ray	length: 0.5 m mass: 7.3 kg wingspan: 1 m	0.4 m/s (0.8 BL/s)	
NTU-RoMan III [61]	manta ray	length: 0.37 m mass: 5 kg wingspan: 0.88 m	0.3 m/s (0.81 BL/s)	
MSU-Grace I [62]	fish	length: 0.5 m mass: 4.2 kg wingspan: 0.32 m volume change: 20 g	0.14 m/s (0.28 BL/s)	
MSU-Grace II [63]	fish	length: 0.9 m mass: 9 kg wingspan: 0.75 m volume change: 100 g	0.2 m/s (0.22 BL/s)	
MSU-Grace III [64]	fish	length: 1.4 m mass: 20 kg wingspan: 0.6 m volume change: 190 g	0.35 m/s (0.25 BL/s)	
Peking-BUG [65]	swordfish	size: 0.89×0.21×0.55 m³ mass: 5.3 kg	–	

motion of manta rays and the low-energy consumption of gliders have been successfully realized. Among them, RoMan II was about 0.5 m long and could glide for 90 hours and flap for 6 hours [60]. RoMan III [61] has reduced the size and made it more compact while maintaining the speed and realized multimodal steering motion, with a minimum steering radius of only 0.01 m. The buoyancy adjustment of the prototypes adopted the suction and drainage mechanism, and the bionic motion drives the flexible fin rays to realize the wave motion based on the sinusoidal waveform [60]. In 2011, the Michigan State University (MSU) proposed a gliding robotic fish [66], which introduced the buoyancy-drive mechanism of the UG into the bionic robotic fish, owning both long endurance and high maneuverability to complete water quality monitoring and other underwater operations. Since 2011, the institution has successively developed three generations of gliding robotic fish (MSU-Grace I, II, and III). Among them, the first generation [62], Tan [66] designed a pair of fixed wings with a large wingspan on the fish-shaped body to generate lift for gliding motion. The gliding robotic fish was equipped with a pump and a water storage cavity to adjust the net buoyancy, and the pitch attitude can be adjusted by moving the battery. However, it had no movable surface to realize the swimming mode and can only perform zigzag gliding motion. The second-generation Grace [63] and the third-generation Grace 2.0 [64] had a rotatable tail fin to achieve helical motion and swimming modes, which can provide the lift required for gliding through a large-sized horizontal fixed surface. Since the tail had only 1 degree of freedom, the swimming performance was weak. Furthermore, harmful algae sampling experiments were conducted in Lake Wintergreen and Lake Higgins in Michigan by combining the spiral gliding motion and body and/or caudal fin swimming of the gliding robotic fish Grace [67]. These gliding robotic fish are equipped with larger wingspans to realize underwater gliding motion while reducing maneuverability [68]. Based on previously mentioned gliding robotic fish, MSU has done a lot of modeling and motion control researches [69]–[74].

Sun Yat-Sen University has also carried out research on the hydrodynamic characteristics of the manta-like underwater glider and designed a conceptual model of the glider with a length of about 83 mm. Through the actual experiments, the results show that the glider will show the optimal gliding performance when the center of mass is 20 mm in front of the center of geometry and the initial attack angle range lies between $-5°$ to $-2.5°$ at the same time [75]. In 2016, the Kyushu University in Japan also conducted some preliminary research on the bionic manta ray glider and carried out bionic shape research through CFD simulation [76].

Zhejiang University has developed an UG with bionic pectoral fins, aiming to explore the effect of pectoral fin aspect ratio on propulsion efficiency. The results show that propulsion efficiency first increases and then decreases with the increase of aspect ratio, and the optimal value is 0.6 to guide the design of the optimal pectoral fin shape [77].

In 2016, the SIA carried out much research on the manta ray–like UG [78], [79], mainly using the high lift–drag ratio shape of the manta ray shape to conduct efficiency optimization work. Besides, the SIA introduced the gliding motion of the UG into the underwater snake robot [59], [80], [81]. Based on the joint design of the bionic robotic snake, telescopic joints were installed at the head and tail to adjust the net buoyancy. A new type of underwater gliding snake-like robot, Perambulator III, was designed. The robot retained the high-mobility swimming gait of the underwater snake-like robot, owning its the long endurance to the simultaneous gliding motion. Compared to UGs, underwater gliding snake-like robots had better maneuverability and were more suitable for applications in shallow waters such as rivers and lakes.

In 2020, Peking University developed a bionic gliding robotic swordfish, named FishBot, which completed the gliding motion and bionic motion by integrating the suction and drainage buoyancy adjustment mechanism of the underwater glider and the multilink robotic fish. During the gliding motion, the diving and rising glide angle were about 7° and 25°, respectively [65].

1.2.4 Thermal UGRs

In 1990, Douglas Webb and Henry Stommel proposed a new scheme for the buoyancy-driving energy of the UG, whose principle was to use the temperature difference between the warm water on the ocean surface and the deep cold water. The theoretical voyage and working time can reach 4×10^4 km and 4 years, respectively [82]. The buoyancy drive unit based on thermal energy mainly consists of a heat exchanger, an accumulator, and a solenoid valve.

One of the key technologies of the thermal UG [83] is to have a hydraulic drive system that can convert the temperature difference energy into mechanical energy. The working principle of the system is shown in Figure 1.4a. There are two kinds of media in the glider: working fluid and transfer fluid. Concretely, the working fluid is sensitive to temperature changes, and a solid–liquid two-state phase transition can occur at room temperature. When the temperature changes from high to low states, the working fluid undergoes a phase change from liquid to solid, leading to the shrink volume [30]. Similarly, when changing from low temperatures to high temperatures, the volume of the working fluid expands. The heat engine system uses the temperature change to expand through the work-ing volume. The volume change caused by the two-phase conversion influ-ences the system buoyancy.

Since 1998, TWR of the United States has successively developed four generations of thermal UGs, aiming to break through the key technology in the ultrahigh range, as shown in Figure 1.4b. However, due to insuffi-cient stability, the developed platform has a high loss rate [30], [82].

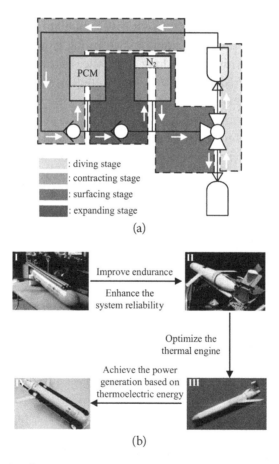

FIGURE 1.4 The illustraation of TUGs. (a) The thermodynamic cycle of TUG [see [30]]. (b) Four generations of Thermal Slocum [see [82]].

Since 2003, Tianjin University has designed a variety of underwater gliders driven by thermal energy [84], [85]. The first-generation system adopted a double-layer structure, which integrated the heat engine with the overall structure of the system to improve the hydrodynamic and motion performance of the underwater gliders. The prototype was tested in the waters of Qiandao Lake in Zhejiang Province, and the effectiveness verification was completed. The second-generation Petrel-II thermal was smaller in size and combined the principle of thermal energy and electrical energy, which avoided the problem of system instability when thermal energy fails [86]. Furthermore, based on the developed platform, Tianjin University has also carried out motion mode optimization and energy consumption analysis [87], [88]. In addition, there was also some research by thermal UG (TUG) at Shanghai Jiaotong University [89], [90].

1.2.5 Other UGRs

There are also some other types of UGR, such as disk-type UGs, blended-wing-body-shape UGs, solar-powered UGs, and so on, as shown in Figure 1.5. Blended-wing-body-shape UGs mainly refer to a type of UG that imitates the shape of the airfoil of an aircraft. The large wingspan makes it have excellent hydrodynamic performance. Since 2003, the Scripps Institution of Oceanography in the United States and the University of Washington have jointly developed large-scale blended-wing-body-shape UGs. The typical representatives were X-Ray (Figure 1.5c) and its improved version Z-Ray (Figure 1.5d), which were similar in shape to manta rays. The gliding motion lift-to-drag ratio can be significantly improved. The Z-RAY had a 6.1 m wingspan and the 17/1 lift-to-drag ratio, which can achieve a horizontal speed of about 1.8 m/s for a 38-L buoyancy engine. The Z-RAY developed in 2010, as shown in Figure 1.5d, had a lift-to-drag ratio of up to 35:1, and can accomplish the task of tracking and identifying marine mammals by airborne acoustic sensors [11]. Northwestern Polytechnical University has carried out a lot of research on airfoil UGs, mainly focusing on its energy consumption analysis and optimization [91], [92], hydrodynamic optimization [93], and gliding motion optimization [94].

Disk-type UGs generally have the shape of a disc and own some unique motion characteristics, such as omnidirectional motion characteristics and ground performance. In 2009, Kyushu University developed a disk-type UG with a length of 1.9 m [95], which was equipped with a suction and drainage buoyancy adjustment mechanism. In particular, four slider

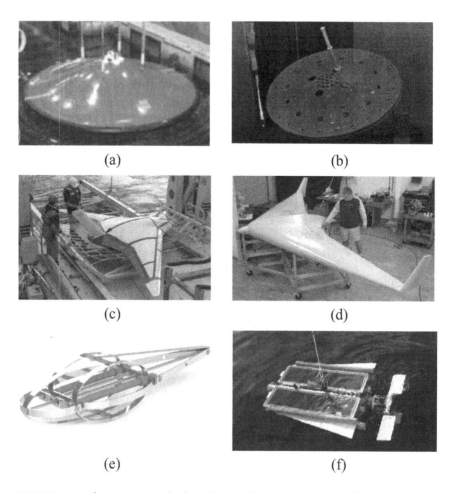

FIGURE 1.5 The overview of other UGs. (a) Boomerang [see [95]]. (b) Disk-type UG [see [96]]. (c) X-Ray [see [11]]. (d) Z-Ray [see [11]]. (e) Anisotropic body UG [see [97]]. (f) Tonai60 [see [98]].

mechanisms were arranged around it to adjust the attitude angle. The UG carried out oceanographic observations in the western waters of Kami-Goto Island, Japan. Dalian Maritime University has also carried out research on disk-type UGs. Aiming at its 3-D dynamic modeling [99], vertical plane control [100], and steering control [101], there were some works. According to the characteristics of large vertical fluid damping of the disk-type underwater glider, a vertical motion method with a zero-pitch angle was proposed, which effectively improved the underwater hovering ability. Regarding the omnidirectional motion characteristics of the circular butterfly, the in situ steering motion method was proposed. In order to improve the hydrodynamic performance and gliding economy of the disk-type UG, a new type of disk-type UG was proposed by Harbin Institute of Technology University. Based on the CFD simulation, the variation law of key parameters at different angles of attack verified the effectiveness of the numerical calculation method [102].

Furthermore, Osaka Prefecture University developed a solar-powered UG, Tonai60 [98], which uses suction and drainage and moving sliders to adjust buoyancy and attitude. The platform was tested in Kagoshima Bay, and the data were successfully recorded using the carried hydroacoustic detector, temperature, and salinity sensor. However, no endurance test was carried out. Samara National Research University proposed a concept of variable shape UG, in which a basic pneumatic circuit was integrated. Depending on the pressure in pneumatic muscles, it can drive the pneumatic muscles fixed to the ribs of the robot's robust body hull deformation [97].

1.3 KEY TECHNOLOGIES OF UGRS

1.3.1 Design of the Buoyancy-Driven System

As the only power source, the buoyancy-driven system is the most important mechanism of the traditional UG, and its performance directly determines the gliding energy consumption and range. In principle, the mission of the buoyancy-driven system is to change the net buoyancy of the body, that is, to change the weight [13] or change the overall volume. Today, the mainstream buoyancy-driven systems adopt the way of changing the volume, which can be divided into three types.

First, the piston-type suction-and-drainage schemes were widely used [19], [26], [32]. In this way, the motor will drive the piston to move in or out to flow the liquid, thereby changing the volume. Second, the two-way-drive internal and external ladder system applies the motor to drive the pump to carry out two-way oil suction. Third, different from the second way, the one-way liquid-system-only drive, the pump transports the liquid from the internal bladder to the external bladder to increase buoyancy [12], [15], [16]. When diving, the liquid can be automatically inhaled with the aid of the atmospheric pressure and the negative pressure of the internal bladder, thereby reducing the buoyancy.

Based on the preceding schemes, from the perspective of the operating depth of the UG, the current buoyancy-driven systems can be classified into two types: shallow sea and deep sea.

1.3.1.1 Shallow-Sea Type

Due to the relatively low pressure in the shallow sea, it can be roughly divided into a piston-type mechanism [26] and an internal and external oil bag system based on a gear pump. Specifically, the advantage of the piston-type mechanism is its high efficiency since it is directly driven by a motor. However, this way is limited by the water depth and motor power; thus, it is only suitable for shallow seas. Most of the current bio-inspired UGs (BUGs) mostly applied this kind of mechanism [62]–[64], [66] and used the syringe method to achieve suction and drainage. However, since the piston needs to be exposed to water, there is a dynamic sealing problem. In addition, the internal and external bladder system based on the gear pump is also another choice. Its advantage lies in the simple structure, but the gear pump has a certain leakage phenomenon.

1.3.1.2 Deepsea Type

The pressure in the deep sea is high, and the axial piston pump is the popular choice. ALACE proposed a detailed buoyancy-driven system scheme based on an axial piston pump, as shown in Figure 1.6 [12]. The system was composed of an outer oil bladder and an inner oil cylinder, and the buoyancy was changed by a one-way oil discharge and negative pressure oil absorption. When the hydraulic oil was all in the inner cylinder, the floating command started the hydraulic pump, leading to the oil entering the outer oil bladder through a one-way valve. The one-way valve can prevent the oil from flowing back to the inner cylinder. When diving, the inner cylinder is in the negative pressure state. Therefore, when opening the solenoid valve at this time, the seawater and atmospheric pressure would press the oil back to the inner cylinder. Besides, the axial piston pump has the advantage of high volumetric efficiency under high pressure, but the oil absorption capacity is insufficient. Since the inner cylinder was in the negative pressure (1/3~1/2 vacuum) environment, tiny vacuum bubbles were easily segregated at the oil inlet end of the micro axial piston pump when transforming the oil from inner to outer bladders. This phenomenon will cause the problem of air lock, and it is difficult to pump the hydraulic oil into the outer bladder.

In response to the airlock phenomenon in ALACE, scientists have successively proposed many solutions. The Seaglider added a booster pump, which sent the hydraulic oil first to the booster pump and then to the high-pressure hydraulic axial piston pump [16]. The designers of Spray believed that the main difficulty with ALACE was the low compression ratio of

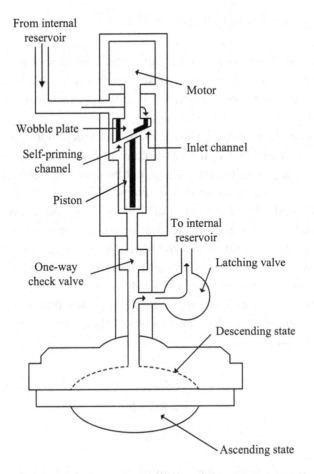

FIGURE 1.6   The buoyancy-driven system of ALACE [see [12]].

the pump [15]. Small air bubbles made the hydraulic oil a compressible fluid, and the piston cannot increase the fluid pressure to the surrounding exhaust pressure due to the limited compression ratio, further stopping the pumping. As a result, Spray's pumps featured a different design with a compression ratio greater than 3:1. There were no ports or hydraulic space due to the use of a pop-up valve. Laboratory tests showed that the pump can move relatively dense slurries of bubbles at high pressure owing to its high compression ratio. Besides, the entrained air bubbles can be effectively directed toward the exhaust valve rather than be trapped. Furthermore, the sea-wing 7000 applied an axial piston pump and a diaphragm liquid pump on the expulsion and inhalation channels [23], respectively, realizing the stable liquid exchange between the inner and outer oil bladders.

1.3.2 System Model of UGRs

1.3.2.1 Dynamic Model

The dynamic model is an important mathematical tool for underwater robot motion simulation and analysis, which can provide reference and guidance for other researchers, such as for performance optimization, state estimation, and control law design.

The dynamic model for gliding motion is crucial for motion analysis and controller design. Regarding the conventional glider, the model is relatively simple since the fuselage will not be bent and deformed. Besides, due to less actuate mechanism, the interaction mode with fluid is single. For traditional UGs, Leonard *et al.* [19] of Princeton University in the United States gave a general method for constructing gliding dynamics equations through momentum and kinetic energy equations. So far, most of the research on UGs took this method as the basis of dynamic analysis. The basic principle of this method is as follows [19].

Denote P and Π as the momentum of the vehicle-fluid system and total angular momentum with respect to (w.r.t.) the body frame. The relationship can be illustrated as

$$p = RP$$
$$\pi = R\Pi + b \times p \qquad (1.1)$$
$$p_p = RP_p,$$

where p and π are the momentum of the vehicle-fluid system and total angular momentum with respect to inertial frame, respectively. b and R are the rotation matrix and position vector of the body frame w.r.t. inertial frame, respectively. P_p represents the point mass momentum with respect to the body frame. Let P_p denote the total momentum of the movable mass $\bar{m}$ w.r.t. the inertial frame. Thus, Newton's laws' state can be derived:

$$\dot{p} = \sum_{i=1}^{I} f_{ext_i}$$
$$\dot{\pi} = \sum_{i=1}^{I} \left(x_i \times f_{ext_i} \right) + \sum_{j=1}^{J} \tau_{ext_j} \qquad (1.2)$$
$$\dot{p}_p = \bar{m}gk + \sum_{k=1}^{K} f_{int_k},$$

where f_{ext} and τ_{ext} indicate the external force and moment, respectively. f_{int} represents the control force. g is gravitational acceleration. k denotes a unit vector in gravity's direction.

Furthermore, the total kinetic energy of the body can be calculated, including the body movement, the attitude adjustment system, and the additional mass. The translational and total angular momentum can be obtained via the derivatives of kinetic energy to translational and angular velocities, respectively. Through differentiating the preceding two different forms of translational and angular momentum, the comprehensive force and moment can be obtained. Then, the dynamic equation can be finally derived through simultaneous solutions. In particular, during the calculation of the resultant external force, the hydrodynamic force is one of the main external forces on the UG. For the dynamic modeling of the UG, the quasi-steady model is generally adopted for analysis. The hydrodynamic forces are related to the velocity and orientation of the robot relative to the current. In order to facilitate analysis, the velocity coordinate system, that is, C_v is often introduced to characterize the relative attitude. Thereafter, the relative orientation is essentially the rotation relation between frame C_w and frame C_b, which can be parameterized by the angle of attack α_b and the sideslip angle β_b.

For the dynamic model of HUG, most of them are based on propeller and buoyancy drive, where buoyancy drive is the same as that of traditional gliders. For propeller drive dynamics, the current modeling methods mainly included the circulation theory design and data mapping methods. However, the circulation theory needs to collect various experimental data of blade sections, the calculation of which is cumbersome, and accurate data need to be supported. Therefore, the general design approach is to use a simple quadratic model combined with empirical data [103]. The propeller thrust f_T and moment τ_T w.r.t. speed n can be described as follows:

$$f_T(\cdot) = sign(n) K_f \rho D^4 n^2 \tag{1.3}$$
$$\tau_T(\cdot) = sign(n) K_\tau \rho D^5 n^2,$$

where K_f and K_τ are positive thrust and moment coefficients, respectively, which can be obtained through an open-water test.

For BUGs, it is indispensable to combine the dynamic models of the UG and robotic fish. Compared with conventional gliders, the modeling complexity of robotic fish is obviously higher. Dynamic modeling of the robotic fish needs to consider three mutually coupled and related parts [104]: hydrodynamics, internal dynamics describing the deformation of the fish body, and external dynamics describing the rigid body motion. The robotic fish realizes various movements through the interaction with the surrounding fluid, and the most difficult problem in the dynamic modeling process lies in the hydrodynamics [104], [105]. The hydrodynamic modeling of the robotic fish often draws on related research on biological fish, such as Lighthill's Elongated Body Theory (EBT) [106] and Large Amplitude Elongated Body Theory (LAEBT) [107]. In addition, the quasi-steady model in airfoil theory [108], [109] was also often applied to describe the hydrodynamic forces on the fins and rigid bodies of the robotic fish. The internal dynamics describe the process of fish body deformation, which needs to be analyzed for the structure and material of the robotic fish. External dynamics describe external motion caused by internal forces, hydrodynamic forces, and other external forces. The methods commonly used in the derivation of the external dynamics model of robotic fish mainly include the Euler–Lagrange [110], Newton–Eulerian [111], Kane [112], and Schiehlen methods [113]. The basic principle and comparison of these methods can be found in Spong [114].

Considering that most of the current BUG adopted rigid body materials, its dynamic modeling can combine the modeling method of traditional UGs with bionic principles. Comparing with the traditional UG, BUG generally controls the attitude via the rotatable surfaces rather than the internal adjustment mechanism. Therefore, rotatable surfaces are not considered in traditional modeling methods. MSU has also carried out three-dimensional dynamic modeling research on the gliding robotic fish, mainly using the tail fin to achieve three-dimensional spiraling motion [72]. However, the previously mentioned studies only established the dynamic model of the gliding motion, lacking the research on bionic motion.

1.3.2.2 Energy Model

For underwater gliding robots, energy consumption is always the most important key indicator. In general, the main energy consumption comes from the buoyancy adjustment system, attitude adjustment system, control system, sensing system, communication system, and propulsion power system.

In terms of the buoyancy drive system, the main difference lies in the electrical and thermal energy. The main energy dissipative components of the electrical buoyancy system method include the solenoid valve and the pump system. Concretely, since the power consumed by the solenoid valve is a fixed value P_v, its energy consumption can be obtained based on the liquid volume ΔV and solenoid flow q_v. In addition, the consumption of the pump system is related to the depth. Generally, there are two ways to calculate the energy. On one hand, the power of the pump at different depths $P_p(z)$ can be measured in advance, and the power-depth function relationship can be fitted to calculate the energy consumption. On the other hand, since the hydraulic pump overcomes the water resistance to accomplish buoyancy change, the energy consumption can be calculated combining the resistance energy with the working efficiency η_p of the hydraulic pump. The final energy consumption, the formula can be described as follows [115], [116]:

$$E_b = E_v + E_p$$

$$E_p = \left\{ \frac{\Delta V}{q_p} P_p(z), \frac{\Delta V \rho g z}{\eta_p} \right\} \qquad (1.4)$$

$$E_v = \frac{\Delta V}{q_v} P_v,$$

where q_p represents the flow.

With regard to the UG based on thermal energy, its energy consumption is mainly from the solenoid valve since the driving force comes from the accumulator and heat exchanger, which consume no more extra energy.

As for the attitude adjustment system, most traditional underwater gliders apply internal moving sliders to adjust the pitch and roll angles, further achieving steering through roll. Generally, it can be assumed that the power of the motor that controls the slider at a fixed speed is a constant value [115], or a function value related to the real-time attitude [116]. Based on this assumption, the energy consumption can be calculated. For bionic underwater gliders, the attitude adjustment system usually adopts a combination of internal sliders and external control surfaces. Concretely, the former is the same as that of UG, and the latter involves complex hydrodynamic analysis with limited studies.

For the control, sensing, and communication systems, their energy consumption is closely related to the working time. The control system generally needs to keep working continuously, and some sensors, such as altimeter, DVL, and so on, work intermittently, while the communication system only works in the stage of out of water. Further, since the working power is less affected by the external environment, their power can basically be assumed to be a fixed constant while the time is variable. For the control system, the total working time is the sum of the underwater gliding time and the surface communication time. The gliding time is geometrically related to the gliding distance, gliding speed, and gliding angle. The surface communication time, including wireless transmission and satellite positioning time, can be set to constant, and the intermittent working time of the sensor can be sampled periodically based on the total duration. Therefore, the variables of the energy consumption are parameters including gliding distance, gliding speed, and gliding angle.

For the propulsion system, the power can be obtained from the propeller dynamics model driven by the hybrid, and the kinetic energy consumption can be derived based on the working time of the propulsion system and the motion efficiency [103]. According to the propeller characteristics, the propeller power consumption can be written as

$$P_a = 2\pi n\tau_T = sign(n)2\pi K_\tau \rho D^5 n^3 \tag{1.5}$$

In terms of the BUG, the power of its propulsion system comes from the fluctuation of the bionic motion mechanism, which generates an interaction force with the external water environment. In theory, the power calculation can also be carried out using the dynamic model, but the difficulty lies in the highly nonlinear characteristics of its hydrodynamics. Compared with the traditional propeller, its parameters are difficult to determine, leading to poor model accuracy, thus there are fewer studies on this problem.

1.3.3 Control of UGRs

1.3.3.1 Motion Control

According to the classification of controlled variables, the related research on gliding motion control can be divided into the categories of vertical motion control, steering control, path tracking, and motion planning. Among them, the objects in the vertical control mainly include pitch angle, glide angle, and depth. The typical control methods are listed in Table 1.4.

Steady-state gliding motion in the vertical plane is one of the most common motions of UGs, resulting in a sawtooth-like motion trajectory.

TABLE 1.4 The Overview of Gliding Motion Control.

Item	Actuator	Method
Pitch control	attitude system [44], [117]–[123], buoyancy system [120],	PID [44], [123], LQR [118]–[120], SMC [117], MPC [121], NNPC [122], backstepping [124], ADRC [125]–[127]
Gliding control	attitude system [19], [128], buoyancy system [19], caudal fin [129], rotatable surfaces [130]	PID [128], [130], LQR [19], nonlinear passive [129]
Depth control	attitude + buoyancy system [131], [132], electrolyzed water [133], [134]	PID [134], [135], LQR [131], [132], state feedback [136],
Yaw control	attitude system [128], tail rudder [44], [137], caudal fin [70] rotatable surfaces [138]	PID [44], [128], SMC [70], perturbation theory [137], backstepping [138]
Path tracking	multiple control unit	SMC [139], MPC [121], [140], H∞ [63], backstepping [141]
Motion planning	-	decentralized supervisory [142], hierarchical supervisory [143]

Generally, the buoyancy adjustment system and the center of gravity adjustment mechanism are applied as the main actuators to change the gliding motion state in the vertical plane.

In aspects of pitch and gliding control, there were various studies. In order to solve the coupling problem of buoyancy adjustment and pitch angle, Yang *et al.* [117] adopted an inverse system approach to decouple the original multiple input multiple output (MIMO) system into two independent single-input, single-output linear subsystems and then controlled each subsystem separately through SMC. The simulation verification was carried out. Considering the effect of water flow disturbance in the glider MIMO system, Isa *et al.* [120] compared the performances of the state feedback controller and linear quadratic regulator (LQR) through simulation analysis. The obtained results showed that the performance of LQR was better than that of the state feedback controller and LQR can compensate for water flow perturbation. Shan *et al.* [121] adopted the Model

Predictive Control (MPC) method based on recurrent neural networks to control the pitch angle, and the effectiveness was validated by simulation. Isa et al. [122] designed a neural network estimation controller (Neural Network Predictive Control, NNPC), and compared it with MPC and LQR through simulation. The simulation results showed that NNPC had better performance in resisting water flow disturbance. Noh et al. designed proportional-integral-derivative (PID) and LQR controllers for underwater gliders, both of which can achieve successful pitch control [118]. Chen et al. presented the LQR controller to control and adjust the vertical motion of the underwater glider [119]. Liu et al. proposed a fuzzy PID controller for the pitch control of the underwater glider, which can adjust the PID parameters online, and verified the effectiveness of methods through lake and sea trials [123]. Isa et al. used a neural network to fit a dynamic model and realized the gliding motion control through MPC [144]. Cao et al. designed an adaptive backstepping controller, whose glide control performance was smoother than that of LQR [124]. Song et al. proposed an active disturbance rejection control (ADRC) based on model step size to improve the control accuracy of pitch angle in gliding motion [125]. Su et al. adopted ADRC for attitude control and modified the control parameters through the reinforcement learning algorithm. The simulation results showed that it had better control performance than the single ADRC controller [126]. Guo et al. proposed an ADRC controller based on model compensation for pitch control in gliding motion, which improved the response speed compared to the pure ADRC method [127]. Leonard et al. [19] established a dynamic model. By linearizing it, an LQR controller was designed to control the glide angle. However, this method assumed that all motion states can be measured and fed back as states. Aiming at the slow convergence speed of traditional PID and LQR controllers, Zhang et al. [129] designed a nonlinear passive controller to control the glide angle. Yan et al. modeled the effect of seawater density and pressure changes on the underwater glider and applied the PID to control the glide angle [130]. In addition to internal sliders, gliders with movable airfoils and bionic gliding gliders have external control surfaces, which can also be used as adjustment mechanisms for two-dimensional gliding motion.

With regard to the depth control of gliding motion, Joo et al. realized the gliding motion by adding the center of gravity adjustment and buoyancy adjustment in the AUV, using the LQR method to control the depth of the zigzag gliding path, and carried out simulation verification [131]. Besides, the authors also investigated the influence of the internal mass on the gliding angle. In recent years, in order to improve the endurance of bionic AUVs, many scholars have also carried out a lot of research on the buoyancy mechanism of bionic robotic fish, most of which were based on traditional mechanisms such as water pumps. In 2014, Makrodimitris *et al.* added a pump-type buoyancy mechanism to the bionic robotic fish and designed a full-state feedback controller to achieve depth control [136]. In 2003, Guo *et al.* proposed a buoyancy regulation scheme based on electrolyzed water and applied it to a micro robotic fish [133]. In 2011, Um *et al.* designed a water electrolysis device using IPMC materials, and the volume of gas generated by electrolysis was controlled by a solenoid valve, thereby changing the device's buoyancy [135]. Subsequently, this device was adopted to achieve deep control with a PID controller, but it is easy to overshoot due to large delay.

In terms of heading control, the traditional UGs achieve steering by a laterally movable mass or a tail rudder. For Slocum, Graver *et al.* [44] designed a PID controller to adjust the yaw angle through the tail rudder and carried out experimental verification. Mahmoudian *et al.* [137] obtained an approximate analytical expression of the steady-state steering motion through the derivation of the canonical perturbation theory, based on which, the required rudder angle was calculated accordingly. Besides, the deviation caused by the approximate processing in the expression can be compensated through feedback control. Zhang *et al.* [70] designed an SMC controller for the gliding robotic fish, achieving directional motion by adjusting the deflection of the tail fin, and verified it experimentally.

The path tracking control has a characteristic of strong comprehensiveness and generally needs to control multiple states at the same time. Since the UG dynamics have significant nonlinear and underactuated features with strong coupling states, path tracking has high requirements for the designed controller. Abraham *et al.* [140] proposed an MPC controller incorporating time-pause technology for the gliding path tracking and compared it with PID through simulation. The simulation results showed that the designed controller had better performance. To track 3-D helical trajectories, Zhang *et al.* [63] designed a two-degree-of-freedom controller consisting of a feed-forward inverse controller and a robust H∞ controller. By comparing with a PI controller pure feedforward inverse controller, the simulation results indicated that the designed controller had better performances in terms of convergence speed, phase delay, and steady-state error.

Motion planning is the control at the decision-making level, whose purpose is to decompose high-level tasks into specific behavioral control flows so that the UG has the ability to make autonomous decisions. Zhang *et al.* [142], [143] proposed a three-layer decentralized supervisory control structure based on the RW (Ramadge and Wonham) supervisory control theory of discrete event dynamical systems. The structure consisted of mission-planning, task-planning, and behavior layers. However, this method was far from practical at present, since the planning layer was difficult to combine with the actual control system at the behavior layer [122].

Although there are many control methods for UGRs to accomplish the underwater task based on the UGs, most of them were for basic motion control. Thus, most of the existing UGRs do not have the ability to operate underwater. By borrowing robotic manipulators followed by their control methods [145], [146], UGRs can obtain the ability in autonomous operation, such as underwater maintenance and handling.

1.3.3.2 Motion Optimization

In order to further increase the range of gliding motion and reduce energy consumption, many scientists have carried out research on the optimization of motion parameters, shape optimization, and optimal motion path of underwater gliding robots.

In terms of motion parameter optimization, Yu *et al.* proposed a motion parameter optimization method and sensor scheduling strategy for the sea-wing underwater glider to improve the range of gliding motion [147]. Xue *et al.* established an energy consumption model for the hybrid-driven underwater glider Petrel-II, and optimized key parameters to reduce the energy consumption under the influence of ocean currents [148]. With the consideration of changes in seawater density, Yang *et al.* improved the range by analyzing gliding strategies and optimizing motion parameters [149]. Song *et al.* established a gliding energy consumption model and a voyage model and analyzed the relationship of gliding motion parameters, energy consumption, and voyage, based on which the optimal parameters were obtained according to the model. The obtained results indicated that the range can be increased by 11.97% after optimization [115]. In addition, Song *et al.* proposed a novel energy model based on the least squares support vector machine (LSSVM) and particle swarm optimization (PSO) algorithm, demonstrating that the LSSVM-PSO model was superior with a large enough training sample size [150]. The previously discussed methods mainly took the maximum range or minimum energy consumption in the steady-state gliding process as the index and calculated the optimal motion parameters of the UG according to the dynamic constraints.

With regard to shape optimization, Sun *et al.* optimized the structural shape of the flying-wing underwater glider to improve the gliding performance [94]. Fu *et al.* proposed a multi-objective shape optimization problem for an underwater glider and solved it using a genetic algorithm to reduce the energy consumption of the gliding motion [151]. Li *et al.* simplified the shape optimization problem by optimizing the cross-sectional airfoil of a flying-wing underwater glider [152]. Dong *et al.* applied a data-driven discrete global optimization algorithm to establish a parametric model of the flying-wing underwater glider structure, further optimizing the buoyancy-to-weight ratio of the cabin-frame coupling structure [153]. Zhang *et al.* improved the Kriging-HDMR method, based on which the lift-drag ratio was improved [154]. Yang *et al.* designed a shape optimization method for the Petrel-L based on the approximate model technology and internal penalty function [155]. The above methods were mainly aimed at the flying wing-shaped underwater glider with a fused wing body, and the relationship between its shape and gliding performance was modeled as a high-dimensional nonlinear problem. Parameters such as lift-drag ratio were taken as the goal to solve the structural shape optimization problem.

As for path planning aiming at optimizing gliding efficiency, Cao *et al.* adopted a genetic algorithm and 3-D Dubins curve to solve the optimal gliding path with the least energy consumption [156]. Liu *et al.* further optimized the 3-D Dubins curve and adjusted the turning radius to improve the range of the underwater glider [157]. The bionic gliding robotic fish has movable pectoral fins that can participate in the regulation of gliding motion. For underwater gliders with independent movable airfoils, there are also some current studies [158], [159].

1.4 DISCUSSION AND FUTURE DEVELOPMENT

1.4.1 Prototype Development

In terms of platform design, traditional underwater gliders have been put into mass production, and their maximum diving depth can reach 10000 meters, which means that most ocean areas can be reached. By carrying various payloads, UGs can effectively explore the ocean. However, the application scenarios of traditional UGs are limited due to poor maneuverability, which can only complete some ocean observation operations. In addition, with the development of technology, various hybrid-driven UGRs, including propeller drive, bionic drive, thermoelectric drive, and the like, have emerged, endowing traditional UGs with stronger motion performances. Specifically, although the propeller-based underwater glider has the characteristics of high swimming speed and strong endurance, it has the disadvantages of low energy efficiency, poor concealment, and damage to the ecological environment. It is not conducive to the sustainable development of the ocean, and it is difficult to adapt to the complex ocean environment near the shallow sea. In recent years, with their unique bionic shape and motion mechanism, bionic underwater gliding robots have gradually attracted the attention of scientific researchers. This type of underwater gliding robot not only has the strong endurance of traditional gliders but also has higher maneuverability than conventional AUVs. However, the current research on the bionic mechanism of BUGs is not sufficient. Due to the limitation of driving materials and mechanisms, its propulsion efficiency is far behind that of biological fish, and further in-depth research is still needed.

Furthermore, it will be toward bionic and intelligent development in the future. On one hand, based on the interdisciplinary characteristics of bionics, robotics, and control, the bionic underwater gliding robot has high maneuverability, strong endurance, strong concealment, and low disturbance by virtue of its streamlined low-resistance shape and unique motion mechanism. It has strong potential in both military and civilian fields. Furthermore, with the development of bionic mechanisms, materials, drivers, and other fields, BUG will play an increasingly important role. On the other hand, with the aid of artificial intelligence, deep learning, reinforcement learning, and other technologies, underwater gliding robots will own the characteristics of strong intelligence. Equipped with a wealth of high-precision sensing systems and computing modules, the underwater gliding robot can obtain intelligent perception, planning, and decision-making capabilities so as to better accomplish the tasks of ocean exploration and resource development. In addition, the group research on underwater gliding robot will also be one of the important fields to enhance their intelligence. Through collaborative cooperation, tasks that a single platform cannot do well can be accomplished.

1.4.2 Technology of Buoyancy-Driven System

With regard to buoyancy-driven system, from the perspective of energy principle, the buoyancy-driven systems of most underwater gliding robots can be divided into electric, thermoelectric, and thermoelectric-electric hybrid driven systems. Among them, the thermoelectric ones mainly use the temperature difference to convert thermal energy into mechanical energy. At present, great progress in TUGs has been made, but the stability and driving ability are still insufficient, leading to application scenarios that are extremely limited. As the mainstream, electric-based buoyancy-driven system mainly includes a piston-type structure for low-pressure environment, and an oil bladder structure for deepwater environment. Most of the piston-type structures are directly driven by motors, while the latter are mostly driven by hydraulic pumps. In general, most UGRs with a maximum diving depth of more than 1000 meters use plunger pumps, which are characterized by high volumetric efficiency under high pressure. Those with a diving depth of fewer than 300 meters mostly use gear pumps or piston drive. On one hand, most of the popular piston-type mechanisms refer to the Slocum shallow-sea glider. On the other hand, most of the deepsea mechanisms are based on ALACE's buoyancy-driven system and solve its airlock phenomenon in various ways. From the point of energy consumption, the shallow-sea drive system consumes energy during both the floating and diving stages. The deep-sea drive system mostly only needs to overcome the pressure in the diving-floating transition stage, while the atmospheric pressure and negative pressure in the cavity are applied to achieve buoyancy reduction in the floating-diving transition stage, so no energy is consumed. However, since the pressure on the water surface is inherently small, the energy saved is still limited, and theoretical breakthroughs in principle are still needed.

In the future, the goal of reducing energy consumption will be committed, and the main research fields may include three aspects. First, on the basis of existing principles, including the development of more efficient hydraulic pumps and motors, some energy consumption should be optimized. Second, some innovations may be urgent. For example, it is well known that water pressure increases with depth, and the main energy consumption of the buoyancy-driven system comes from the high-pressure liquid discharge operation at the diving to surfacing point. Therefore, how to use the natural characteristics of water pressure to improve the buoyancy-driven system is also a question worth exploring. Third, the performance based on TUG may be optimized. The existing thermoelectric energy system has poor stability, and its application is much limited in the operating area. Therefore, how to develop new materials and optimize the driving mechanism needs to be further explored.

1.4.3 Motion Control and Optimization

Regarding the motion control and optimization, most of UGRs currently focus on system modeling, attitude control, depth control, path tracking and planning. Since underwater gliding robots commonly adopt rigid shells and their motion mechanisms, their motion and energy consumption modeling is not complicated. Comparatively speaking, it is relatively difficult to establish the system model for BUGs due to the bionic motion mechanism. As the most difficult part of motion modeling, hydrodynamic is generally conducted by a quasi-steady-state lift-drag model. However, the ocean environment is extremely complicated owing to unpredictable ocean currents and interference. As can be seen in Table 1.4, the current motion control algorithms often applied model-free methods, such as PID, fuzzy, and so on, but these algorithms are easily affected by external interference with poor robustness. Therefore, some model-based robust control algorithms have received attention in recent years. Some methods do not require accurate models, such as sliding mode control and model predictive control. However, the implementation of the above methods may be slightly more difficult, and the control parameters are also more complicated. Besides, since the underwater gliding robot has the obvious characteristics of time delay during the gliding motion, the above methods are not suitable to apply directly due to poor convergence, which usually needs to be improved.

Moreover, with the rapid development of artificial intelligence algorithms, future control algorithms will be enhanced in terms of strong intelligence, adaptability, and anti-disturbance. With the aid of deep reinforcement learning algorithm, the intelligent control of robots has attracted the attention of researchers. Through the preliminary offline training, UGRs can obtain strong environmental adaptability, so as to better complete ocean operations. However, the ocean environment is extremely complex, and the disturbance of ocean currents is unpredictable. Therefore, how to ensure the convergence of the learning algorithm and optimization methods, further improving the stability of the system is a problem worth exploring. Besides, in order to improve the system robustness, some optimization-based random searching algorithms, such as bio-inspired control and neural network, are also effective solutions [160]–[162]. In addition, accurate perception ability is an important basis for control tasks. With the consideration of poor underwater visual environments, such as weak light, strong attenuation, and blurring, the requirements of perception systems and algorithms are also relatively high. Besides, conventional acoustic equipment has a relatively long detection range, but lacks perceptual details and intelligence, while traditional visual equipment is greatly affected by the water quality environment owing to poor image quality and a small field of view. Therefore, how to fuse the information of multiple sensors to provide high-quality input for motion control needs to be further explored. Besides, since the underwater medium is different from air, it is necessary to learn from the calibration algorithm of industrial robots to achieve accurate underwater visual calibration [163]. Furthermore, underwater simultaneous localization and mapping is also another hot issue. Under sensor information which may be unreliable and sparse, some intelligent methods via referring to the field of industrial robots and applications need to be explored [164], [165], further building the accurate map.

1.4.4 Application Scenarios Prospect of UGRs

Most current applications of underwater gliders focused on ocean observation missions. According to different observation scenarios, it can be mainly divided into ocean hydrological observation, profile observation of seabed topography, biological observation of marine scientific exploration, and so on. Through the onboard payloads, such as water quality sensors, conductance, temperature and depth sensors, acoustic sensors, hydrophones, current meters, and turbulence sensors, real-time underwater information can be collected, stored, and sent to shore-based systems through satellite or wireless communication. For example, Wall *et al.* integrated a hydrophone into the UG, and successfully detected the sound of biological fish in a large space [166]. Schultze *et al.* acquired ocean microstructure turbulence data sets using UG and analyzed the dissipation rate of ocean turbulent kinetic energy [167]. However, today, due to the poor maneuverability of UGs, it is difficult to complete underwater tasks in complex sea areas, hindering its application development.

With the development of ocean science and robotics, UGRs play an increasingly important role in various ocean applications. With the emergence of new UGRs, the motion performances have been strengthened, greatly expanding the application field:

- The establishment of an ocean stereoscopic observation network will be an important application scenario in the future. Ocean science needs the support of long-term observation and depends on data accumulation of the ocean. The UGRs can not only achieve long-term all-weather observation but also ensure its high observation accuracy and resolution due to its high maneuverability.

- Deep-sea archaeology and rescue are another typical point. Most of the existing technologies rely on ROVs with limited operating range, while traditional AUVs cannot guarantee long-lasting operations. Therefore, by carrying payloads, UGRs will play an important role in the future.

- With consideration of the complex environment in the shallow sea, it is difficult to meet the task requirements for the traditional UGs due to their poor maneuverability. The new UGRs, especially the BUGs, can not only realize the near-shallow sea operation, such as topographic and landform observation but also will not cause harm to the environment.

1.5 CONCLUDING REMARKS

Conquering the vast ocean has always been an urgent pursuit of human beings. The diversity of underwater vehicles for ocean exploration has been enriched. To overcome the endurance of underwater vehicles, UGRs have been widely investigated and increasingly deployed in recent decades. Further, in view of its existing issues of designed prototypes, many new type platforms are gradually being developed to better accomplish ocean exploration. In this chapter, comprehensive research into the underwater gliding robots has been conducted in terms of state-of-the-art prototypes and their key technologies. More specifically, the traditional UGs, hybrid-driven UGs, bio-inspired UGs, thermal UGs, and other types are introduced in detail. Furthermore, the key technologies, involving the buoyancy-driven system, dynamic and energy models, and motion control and optimization are analyzed and synthesized. Finally, the critical issues and future development are discussed. Thanks to the rapid development of artificial intelligence, communication technology, and internet of things, it is believed that UGRs will substantially contribute to underwater operations, further unfolding an enormous potential in sustainable exploration and exploitation of oceans.

REFERENCES

[1] G. N. Roberts and R. Sutton, "Advances in unmanned marine vehicles," *IET*, vol. 69, 2006.
[2] C. Tang, Y. Wang, S. Wang, R. Wang, and M. Tan, "Floating autonomous manipulation of the underwater biomimetic vehicle-manipulator system: Methodology and verification," *IEEE Trans. Ind. Electron.*, vol. 65, no. 6, pp. 4861–4870, 2017.

[3] E. Fiorelli, N. E. Leonard, P. Bhatta, D. A. Paley, R. Bachmayer, and D. M. Fratantoni, "Multi-AUV control and adaptive sampling in Monterey Bay," *IEEE J. Ocean Eng.*, vol. 31, no. 4, pp. 935–948, 2006.

[4] L. Paull, S. Saeedi, M. Seto, and H. Li, "AUV navigation and localization: A review," *IEEE J. Ocean Eng.*, vol. 39, no. 1, pp. 131–149, 2013.

[5] U. K. Verfuss et al., "A review of unmanned vehicles for the detection and monitoring of marine fauna," *Mar. Pollut. Bull.*, vol. 149, pp. 17–29, 2019.

[6] D. L. Rudnick, "Ocean research enabled by underwater gliders," *Annu. Rev. Mar. Sci.*, vol. 8, pp. 519–541, 2016.

[7] R. E. Davis, C. E. Eriksen, and C. P. Jones, "Autonomous buoyancy-driven underwater gliders," *Technology and Applications of Autonomous Underwater Vehicles*, vol. 3, pp. 37–58, 2002.

[8] D. L. Rudnick, R. E. Davis, C. C. Eriksen, D. M. Fratantoni, and M. J. Perry, "Underwater gliders for ocean research," *Mar. Technol. Soc. J.*, vol. 38, no. 2, pp. 73–84, 2004.

[9] M. Y. Javaid, M. Ovinis, T. Nagarajan, and F. B. Hashim, "Underwater gliders: A review," *EDP Sci.*, vol. 13, 2014.

[10] B. Ullah, M. Ovinis, M. B. Baharom, M. Y. Javaid, and S. S. Izhar, "Underwater gliders control strategies: A review," in *Proc. Asian Control Conf.*, Malaysia, 2015, pp. 1–6.

[11] X. Shen, Y. Wang, S. Yang, Y. Yan, and H. Li, "Development of underwater gliders: An overview and prospect," *J. Unmanned Undersea Syst.*, vol. 26, pp. 89–106, 2018.

[12] R. E. Davis, D. C. Webb, L. A. Regier, and J. Dufour, "The autonomous lagrangian circulation explorer (ALACE)," *J. Atmos Ocean. Technol.*, vol. 9, pp. 264–285, 1992.

[13] K. Kawaguchi, T. Ura, Y. Tomoda, and H. Kobayashi, "Development and sea trials of a shuttle type AUV 'ALBAC'," in *Proc. Int. Symp. Unmanned Untethered Submersible Technol.*, University of New Hampshire-Marine Systems, 1993, pp. 7–13.

[14] T. J. Osse and C. C. Eriksen, "The deepglider: A full ocean depth glider for oceanographic research," in *Proc. OCEANS*, BC, Canada, 2007, pp. 1–12.

[15] J. Sherman, R. E. Davis, W. B. Owens, and J. Valdes, "The autonomous underwater glider 'Spray'," *IEEE J. Ocean Eng.*, vol. 26, no. 4, pp. 437–446, 2001.

[16] C. C. Eriksen, et al., "Seaglider: A long-range autonomous underwater vehicle for oceanographic research," *IEEE J. Ocean Eng.*, vol. 26, no. 4, pp. 424–436, 2001.

[17] O. Schofield, et al., "Slocum gliders: Robust and ready," *J. Field Robot*, vol. 24, no. 6, pp. 473–485, 2007.

[18] H. Claustre, L. Beguery, and P. Pla, "SeaExplorer glider breaks two world records multisensor UUV achieves global milestones for endurance, distance," *Sea Technol.*, vol. 55, no. 3, pp. 19–21, 2014.

[19] N. E. Leonard and J. G. Graver, "Model-based feedback control of autonomous underwater gliders," *IEEE J. Ocean Eng.*, vol. 26, no. 4, pp. 633–645, 2001.

[20] S. Yamaguchi, T. Naito, T. Kugimiya, K. Akahoshi, and M. Fujimoto, "Development of a motion control system for underwater gliding vehicle," in *Proc. Int. Offshore Polar Eng. Conf.*, Lisbon, Portugal, 2007, pp. 1115–1120.

[21] N. Ichihashi, T. Ikebuchi, and M. Arima, "Development of an underwater glider with independently controllable main wings," in *Proc. Int. Offshore Polar Eng. Conf.*, BC, Canada, 2008, pp. 156–161.

[22] J. Yu, A. Zhang, W. Jin, Q. Chen, Y. Tian, and C. Liu, "Development and experiments of the sea-wing underwater glider," *China Ocean Eng.*, vol. 25, no. 4, pp. 721–736, 2011.

[23] J. Yu, W. Jin, Z. Tan, Y. Huang, Y. Luo, and X. Wang, "Development and experiments of the Sea-Wing7000 underwater glider," in *Proc. OCEANS*, AK, USA, 2017, pp. 1–7.

[24] D. L. Rudnick, R. E. Davis, and J. T. Sherman, "Spray underwater glider operation," *J. Atmos. Ocean Tech.*, vol. 33, no. 6, pp. 1113–1122, 2016.

[25] C. Jones et al., *Slocum Glider Expanding the Capabilities*. Portsmouth, 2011.

[26] C. Jones, B. Allsup, and C. DeCollibus, "Slocum Glider: Expanding our understanding of the oceans," in *Proc. IEEE OCEANS*, St. John's, 2014, pp. 14–19.

[27] P. Pla and R. Tricarico, "Towards a low cost observing system based on low logistic SeaExplorer glider," in *Proc. IEEE Underwater Technol.*, Chennai, India, 2015, pp. 1–3.

[28] O. D. Fommervault, F. Besson, L. Beguery, Y. L. Page, and P. Lattes, "Seaexplorer underwater glider: A new tool to measure depth-resolved water currents profiles," in *Proc. IEEE OCEANS*, Marseille, France, 2019, pp. 1–6.

[29] H. Klinck et al., "Near-real-time acoustic monitoring of beaked whales and other cetaceans using a Seaglider™," *PLoS One*, vol. 7, p. 5, 2012.

[30] D. C. Webb, P. J. Simonetti, and C. P. Jones, "SLOCUM: An underwater glider propelled by environmental energy," *IEEE J. Ocean Eng.*, vol. 26, no. 4, pp. 447–452, 2001.

[31] A. Alvarez, "Redesigning the SLOCUM glider for torpedo tube launching," *IEEE J. Ocean Eng.*, vol. 35, no. 4, pp. 984–991, 2010.

[32] J. Yu, Q. Zhang, L. Wu, and A. Zhang, "Movement mechanism design and motion performance analysis of an underwater glider," *Robot*, vol. 27, no. 5, pp. 390–395, 2005.

[33] *Slocum G3 glider autonomous underwater vehicle long endurance, proven performance* [Online]. Available: http://obsplatforms.plocan.eu/media/datasheets

[34] A. Caffaz, A. Caiti, G. Casalino, and A. Turetta, "The hybrid glider/AUV fòlaga," *IEEE Robot Autom. Mag.*, vol. 17, no. 1, pp. 31–44, 2010.

[35] J. Sliwka, B. Clement, and I. Probst, "Sea glider guidance around a circle using distance measurements to a drifting acoustic source," in *Proc. IEEE/RSJ Int. Conf. Intell. Robots Syst.*, Portugal, 2012, pp. 94–99.

[36] Y. Huang, J. Qiao, J. Yu, Z. Wang, Z. Xie, and K. Liu, "Sea-Whale 2000: A long-range hybrid autonomous underwater vehicle for ocean observation," in *Proc. IEEE OCEANS*, Marseille, France, 2019, pp. 1–6.

[37] Z. Chen, J. Yu, A. Zhang, and F. Zhang, "Design and analysis of folding propulsion mechanism for hybrid-driven underwater gliders," *Ocean Eng.*, vol. 119, pp. 125–134, 2016.

[38] S. Wang, X. Sun, J. Wu, X. Wang, and H. Zhang, "Motion characteristic analysis of a hybrid-driven underwater glider," in *Proc. OCEANS*, NSW, Australia, 2010, pp. 1–9.

[39] S. Wang, H. Li, Y. Wang, Y. Liu, H. Zhang, and S. Yang, "Dynamic modeling and motion analysis for a dual-buoyancy-driven full ocean depth glider," *Ocean Eng.*, vol. 187, 2019.

[40] J. Cao, D. Lu, D. Li, Z. Zeng, B. Yao, and L. Lian, "Smartfloat: A multimodal underwater vehicle combining float and glider capabilities," *IEEE Access Pract. Innov. Open Solut.*, vol. 7, pp. 77825–77838, 2019.

[41] A. Alvarez et al., "Design and realization of a very low cost prototypal autonomous vehicle for coastal oceanographic missions," *IFAC Proc.*, vol. 37, no. 10, pp. 471–476, 2004.

[42] A. Alvarez et al., "Fòlaga: A very low cost autonomous underwater vehicle for coastal oceanography," *IFAC Proc.*, vol. 38, no. 1, pp. 31–36, 2005.

[43] A. Alvarez et al., "Fòlaga: A low-cost autonomous underwater vehicle combining glider and AUV capabilities," *Ocean Eng.*, vol. 36, no. 1, pp. 24–38, 2009.

[44] J. G. Graver, "Underwater gliders: Dynamics, control and design," Princeton University, 2005.

[45] Y. Huang, Z. Wang, J. Yu, A. Zhang, J. Qiao, and H. Feng, "Development and experiments of the passive buoyancy balance system for sea-whale 2000 AUV," in *Proc. IEEE OCEANS*, Marseille, France, 2019, pp. 1–5.

[46] Z. Z. Chen, J. Yu, A. Zhang, R. Yi, and Q. Zhang, "Folding propeller design and analysis for a hybrid driven underwater glider," in *Proc. OCEANS*, CA, USA, 2013, pp. 1–9.

[47] Z. Chen, "Research on system efficiency and motion modeling problems for hybrid-driven underwater gliders," University of Chinese Academy of Sciences, 2016.

[48] W. Niu, S. Wang, Y. Wang, Y. Song, and Y. Zhu, "Stability analysis of hybrid-driven underwater glider," *China Ocean Eng.*, vol. 31, no. 5, pp. 528–538, 2017.

[49] F. Liu, "System design and motion behaviors analysis of the hybrid underwater glider," Tianjin University, 2014.

[50] D. Lu et al., "Design, fabrication, and characterization of a multimodal hybrid aerial underwater vehicle," *Ocean Eng.*, vol. 219, 2021.

[51] J. Cao, D. Li, Z. Zeng, B. Yao, and L. Lian, "Drifting and gliding: Design of a multimodal underwater vehicle," in *Proc. IEEE OCEANS*, Kobe, Japan, 2018, pp. 1–7.

[52] D. Lu, C. Xiong, Z. Zeng, and L. Lian, "A multimodal aerial underwater vehicle with extended endurance and capabilities," in *Proc. IEEE Int. Conf. Rob. Autom.*, Montreal, QC, Canada, 2019, pp. 4674–4680.

[53] H. Zhou et al., "Dynamic modeling and motion control of a novel conceptual multimodal underwater vehicle for autonomous sampling," *Ocean Eng.*, vol. 240, 2021.

[54] K. Alam, T. Ray, and S. G. Anavatti, "Design optimization of an unmanned underwater vehicle using low- and high-fidelity models," *IEEE Trans. Syst. Man Cybern. Syst.*, vol. 47, no. 11, pp. 2794–2808, 2017.

[55] R. K. Katzschmann, J. DelPreto, R. MacCurdy, and D. Rus, "Exploration of underwater life with an acoustically controlled soft robotic fish," *Sci. Robot.*, vol. 3, p. 16, 2018.

[56] R. Wang, S. Wang, Y. Wang, M. Tan, and J. Yu, "A paradigm for path following control of a ribbon-fin propelled biomimetic underwater vehicle," *IEEE Trans. Syst. Man Cybern. Syst.*, vol. 49, no. 3, pp. 482–493, 2019.

[57] J. Yu, Z. Su, M. Wang, M. Tan, and J. Zhang, "Control of yaw and pitch maneuvers of a multilink dolphin robot," *IEEE Trans. Robot.*, vol. 28, no. 2, pp. 318–329, 2012.

[58] J. Yu, Z. Su, Z. Wu, and M. Tan, "An integrative control method for bio-inspired dolphin leaping: Design and experiments," *IEEE Trans. Ind. Electron.*, vol. 63, no. 5, pp. 3108–3116, 2016.

[59] J. Tang, B. Li, Z. Li, and J. Chang, "A novel underwater snake-like robot with gliding gait," in *Proc. IEEE Int. Conf. CYBER Technol. Autom., Control, Intell. Syst.*, Honolulu, HI, USA, 2017, pp. 1113–1118.

[60] C. Zhou and K. H. Low, "Better endurance and load capacity: An improved design of manta ray robot (RoMan-II)," *J. Bionic. Eng.*, vol. 7, pp. S137–S144, 2010.

[61] K. H. Low, C. Zhou, G. Seet, S. Bi, and Y. Cai, "Improvement and testing of a robotic manta ray (RoMan-III)," in *Proc. IEEE Int. Conf. Robot. Biomim.*, Karon Beach, Thailand, 2011, pp. 1730–1735.

[62] F. Zhang, J. Thon, C. Thon, and X. Tan, "Miniature underwater glider: Design and experimental results," *IEEE/ASME Trans. Mechatronics*, vol. 19, no. 1, pp. 394–399, 2013.

[63] F. Zhang and X. Tan, "Three-dimensional spiral tracking control for gliding robotic fish," in *Proc. IEEE Conf. Decis. Control*, California, USA, 2014, pp. 5340–5345.

[64] O. Ennasr, G. Mamakoukas, T. Murphey, and X. Tan, "Ergodic exploration for adaptive sampling of water columns using gliding robotic fish," Atlanta, Georgia, USA pp. 51913, 2018.

[65] C. Wang, J. Lu, X. Ding, C. Jiang, J. Yang, and J. Shen, "Design, modeling, control, and experiments for a fish-robot-based IoT platform to enable smart ocean," *IEEE Internet Things J.*, vol. 8, no. 11, pp. 9317–9329, 2021.

[66] X. Tan, "Autonomous robotic fish as mobile sensor platforms: Challenges and potential solutions," *Mar. Technol. Soc. J.*, vol. 45, no. 4, pp. 31–40, 2011.

[67] F. Zhang, O. Ennasr, E. Litchman, and X. Tan, "Autonomous sampling of water columns using gliding robotic fish: Algorithms and harmful-algae-sampling experiments," *IEEE Syst. J.*, vol. 10, no. 3, pp. 1271–1281, 2015.

[68] O. N. Ennasr, *Gliding Robotic Fish: Design, Collaborative Estimation, and Application to Underwater Sensing.* Michigan State University, 2020.

[69] F. Zhang, X. Tan, and H. K. Khalil, "Passivity-based controller design for stablization of underwater gliders," in *Proc. Am. Control Conf.*, Canada, 2012, pp. 5408–5413.

[70] F. Zhang and X. Tan, "Gliding robotic fish and its tail-enabled yaw motion stabilization using sliding mode control," Palo Alto, California, USA pp. 56130, 2013.

[71] F. Zhang and X. Tan, "Nonlinear observer design for stabilization of gliding robotic fish," in *Proc. Am. Control Conf.*, Oregon, USA, 2014, pp. 4715–4720.

[72] F. Zhang, F. Zhang, and X. Tan, "Tail-enabled spiraling maneuver for gliding robotic fish," *J. Dyn. Syst. Meas. Control*, vol. 136, p. 4, 2014.

[73] F. Zhang, O. Ennasr, and X. Tan, "Gliding robotic fish: An underwater sensing platform and its spiral-based tracking in 3D space," *Mar. Technol. Soc. J.*, vol. 51, no. 5, pp. 71–78, 2017.

[74] D. Coleman and X. Tan, "Backstepping control of gliding robotic fish for trajectory tracking in 3D space," in *Proc. Am. Control Conf.*, CO, USA, 2020, pp. 3730–3736.

[75] W. Cai, J. Zhan, and Y. Luo, "A study on the hydrodynamic performance of manta ray biomimetic glider under unconstrained six-DOF motion," *PLoS One*, vol. 15, p. 11, 2020.

[76] S. Yamaguchi, H. Mizunaga, T. Katsu, S. Nakamuta, and Y. Kono, "Preliminary design of an underwater glider for ocean floor resources exploration," in *Proc. Int. Offshore Polar Eng. Conf.*, Rhodes, Greece, 2016, pp. 590–594.

[77] Y. Li, D. Pan, Z. Ma, and Q. Zhao, "Aspect ratio effect of a pair of flapping wings on the propulsion of a bionic autonomous underwater glider," *J. Bionic Eng.*, vol. 16, pp. 145–153, 2019.

[78] Z. Wang, J. Yu, and A. Zhang, "Hydrodynamic performance analysis of a biomimetic manta ray underwater glider," in *Proc. IEEE Int. Conf. Robot. Biomim.*, Qingdao, China, 2016, pp. 1631–1636.

[79] Z. Wang, J. Yu, A. Zhang, Y. Wang, and W. Zhao, "Parametric geometric model and hydrodynamic shape optimization of a flying-wing structure underwater glider," *China Ocean Eng.*, vol. 31, no. 6, pp. 709–715, 2017.

[80] X. Zhang, B. Li, J. Chang, and J. Tang, "Gliding control of underwater gliding snake-like robot based on reinforcement learning," in *Proc. IEEE Int. Conf. CYBER Technol. Autom., Control, Intell. Syst.*, Tianjin, China, 2018, pp. 323–328.

[81] J. Tang, B. Li, J. Chang, and A. Zhang, "Unscented Kalman-filter-based sliding mode control for an underwater gliding snake-like robot," *Sci. China Inf. Sci.*, vol. 2207, no. 7, pp. 1–11220, 2020.

[82] Y. Yang, "System design and performance analysis of thermal-electric hybrid power underwater glider," Tianjin University, 2017.

[83] Y. Wang, S. Wang, and C. Xie, "Dynamic analysis and system design on an underwater glider propelled by temperature difference energy," *J. Tianjin Univ.*, vol. 40, no. 2, pp. 133–138, 2007.

[84] H. Zhang, Y. Wang, and Z. Lian, "Application and improvement of the interlayer thermal engine powered by ocean thermal energy in an underwater glider," in *Proc. IEEE Asia-Pacific Pow. Energy Eng. Conf.*, Wuhan, China, 2009, pp. 1–4.

[85] G. Wang, Y. Yang, and S. Wang, "Ocean thermal energy application technologies for unmanned underwater vehicles: A comprehensive review," *Appl. Energy*, vol. 278, 2020.

[86] Y. Yang, Y. Wang, Z. Ma, and S. Wang, "A thermal engine for underwater glider driven by ocean thermal energy," *Appl. Therm. Eng.*, vol. 99, pp. 455–464, 2016.

[87] Z. Ma, Y. Liu, Y. Wang, and S. Wang, "Improvement of working pattern for thermal underwater glider," in *Proc. IEEE OCEANS*, Shanghai, China, 2016, pp. 1–6.

[88] B. Li, Y. Yang, L. Zhang, and S. Wang, "Research on sailing range of thermal-electric hybrid propulsion underwater glider and comparative sea trial based on energy consumption," *Appl. Ocean Res.*, vol. 114, 2021.

[89] Y. Hai and M. Jie, "Optimization of displacement and gliding path and improvement of performance for an underwater thermal glider," *J. Hydrodyn.*, vol. 22, no. 5, pp. 618–625, 2010.

[90] Q. Kong, J. Ma, and D. Xia, "Numerical and experimental study of the phase change process for underwater glider propelled by ocean thermal energy," *Renew. Energy*, vol. 35, no. 4, pp. 771–779, 2010.

[91] C. Sun, B. Song, P. Wang, and B. Zhang, "Energy consumption optimization of steady-state gliding for a Blended-Wing-Body underwater glider," in *Proc. OCEANS*, CA, USA, 2016, pp. 1–5.

[92] X. Du and L. Zhang, "Analysis on energy consumption of blended-wing-body underwater glider," *Int. J. Adv. Rob. Syst.*, vol. 17, no. 2, pp. 1–9, 2020.

[93] D. Lyu, B. Song, G. Pan, Z. Yuan, and J. Li, "Winglet effect on hydrodynamic performance and trajectory of a blended-wing-body underwater glider," *Ocean Eng.*, vol. 188, 2019.

[94] C. Sun, B. Song, P. Wang, and X. Wang, "Shape optimization of blended-wing-body underwater glider by using gliding range as the optimization target," *Int. J. Nav. Arch. Ocean Eng.*, vol. 9, no. 6, pp. 693–704, 2017.

[95] M. Nakamura et al., "Disk type underwater glider for virtual mooring and field experiment," *Int. Offshore Polar Eng.*, vol. 19, no. 1, pp. 66–70, 2009.

[96] H. Zhou, "Research on motion characteristics of the disk-type underwater glider," Dalian Maritime University, 2021.

[97] I. D. Galushko, V. A. Salmina, S. A. Gafurov, and D. M. Stadnik, "Approach of flow control around unmanned underwater robot," *IFAC-PaperOnLine*, vol. 51, no. 30, pp. 452–457, 2018.

[98] M. Arima, H. Tonai, and K. Yoshida, "Development of an ocean-going solar-powered underwater glider," in *Proc. Int. Offshore Polar Eng. Conf.*, Busan, Korea, 2014, pp. 444–448.

[99] P. Yu, T. Wang, H. Zhou, and C. Shen, "Dynamic modeling and three-dimensional motion simulation of a disk type underwater glider," *Int. J. Nav. Arch. Ocean Eng.*, vol. 10, no. 3, pp. 318–328, 2018.

[100] H. Zhou, T. Wang, L. Sun, and W. Lan, "Study on the vertical motion characteristics of disc-type underwater gliders with zero pitch angle," *J. Mar. Sci. Technol.*, vol. 25, no. 3, pp. 828–841, 2020.

[101] H. Zhou, T. Wang, L. Sun, and X. Jin, "Disc-type underwater glider modeling and analysis for omnidirectional and steering motion characteristics," *Int. J. Control Autom. Syst.*, vol. 19, no. 1, pp. 532–547, 2021.

[102] J. Wang, C. Wang, Y. Wei, and C. Zhang, "Hydrodynamic characteristics and motion simulation of flying-wing dish-shaped autonomous underwater glider," *J. Harbin Inst. Technol.*, vol. 50, pp. 131–137, 2018.

[103] A. J. Sørensen and Ø. N. Smogeli, "Torque and power control of electrically driven marine propellers," *Control Eng. Pract.*, vol. 17, no. 9, pp. 1053–1064, Sep. 2009, doi: 10.1016/j.conengprac.2009.04.006.

[104] F. Boyer, M. Porez, A. Leroyer, and M. Visonneau, "Fast dynamics of an eel-like robot-comparisons with Navier-Stokes simulations," *IEEE Trans. Robot.*, vol. 24, no. 6, pp. 1274–1288, 2008.

[105] J. E. Colgate and K. M. Lynch, "Mechanics and control of swimming: A review," *IEEE J. Ocean Eng.*, vol. 29, no. 3, pp. 660–673, 2004.

[106] M. J. Lighthill, "Note on the swimming of slender fish," *J. Fluid Mech.*, vol. 9, no. 2, pp. 305–317, 1960.

[107] M. J. Lighthill, "Large-amplitude elongated-body theory of fish locomotion," *R. Soc. Lond.*, vol. 179, no. 1055, pp. 125–138, 1971.

[108] M. Aureli, V. Kopman, and M. Porfiri, "Free-locomotion of underwater vehicles actuated by ionic polymer metal composites," *IEEE/ASME Trans. Mechatronics*, vol. 15, no. 4, pp. 603–614, 2010.

[109] V. Kopman, J. Laut, F. Acquaviva, and A. Rizzo, "Dynamic modeling of a robotic fish propelled by a compliant tail," *IEEE J. Ocean Eng.*, vol. 40, no. 1, pp. 209–221, 2015.

[110] D. H. El, T. Salumae, L. D. Chambers, W. M. Megill, and M. Kruusmaa, "Modelling of a biologically inspired robotic fish driven by compliant parts," *Bioinspir. Biomim.*, vol. 9, p. 1, 2014.

[111] W. Khalil, G. Gallot, and F. Boyer, "Dynamic modeling and simulation of a 3-D serial eel-like robot," *IEEE Trans. Syst. Man Cybern. Cybern.*, vol. 37, no. 6, pp. 1259–1268, 2007.

[112] Z. Wu, J. Yu, and M. Tan, "CPG parameter search for a biomimetic robotic fish based on particle swarm optimization," in *Proc. IEEE Int. Conf. Robot. Biomimet. (ROBIO)*, Guangzhou, China, 2012, pp. 563–568.

[113] J. Yu, L. Liu, and M. Tan, "Dynamic modeling of multi-link swimming robot capable of 3-D motion," in *Proc. IEEE Int. Conf. Mechatronics Autom.*, Harbin, China, 2007, pp. 1322–1327.

[114] M. W. Spong, *Robot Dynamics and Control*. John and Sons; Wiley, 1989.

[115] Y. Song, Y. Wang, S. Yang, S. Wang, and M. Yang, "Sensitivity analysis and parameter optimization of energy consumption for underwater gliders," *Energy*, vol. 191, 2020.

[116] Y. Song, H. Ye, Y. Wang, W. Niu, X. Wan, and W. Ma, "Energy consumption modeling for underwater gliders considering ocean currents and seawater density variation," *J. Mar. Sci. Eng.*, vol. 9, p. 11, 2021.

[117] H. Yang and J. Ma, "Nonlinear control for autonomous underwater glider motion based on inverse system method," *J. Shanghai Jiaotong Univ.*, vol. 15, no. 6, pp. 713–718, 2010.

[118] M. M. Noh, M. R. Arshad, and R. M. Mokhtar, "Depth and pitch control of USM underwater glider: Performance comparison PID vs. LQR," *Indian J. Geo-Mar. Sci.*, vol. 40, no. 2, pp. 200–206, 2011.

[119] Y. Chen, W. Yan, J. Gao, and L. Du, "Modeling and vertical motion control of underwater glider," *Fire Control Command Control*, vol. 37, no. 4, pp. 141–145, 2012.

[120] K. Isa and M. Arshad, "Buoyancy-driven underwater glider modelling and analysis of motion control," *Indian J. Geo–Marine Sci.*, vol. 41, no. 6, pp. 516–526, 2012.

[121] Y. Shan, Z. Yan, and J. Wang, "Model predictive control of underwater gliders based on a one-layer recurrent neural network," in *Proc. IEEE Int. Conf. Adv. Comput. Intell.*, Hangzhou, China, 2013, pp. 328–333.

[122] K. Isa, M. Arshad, and S. Ishak, "A hybrid-driven underwater glider model, hydrodynamics estimation, and an analysis of the motion control," *Ocean Eng.*, vol. 81, pp. 111–129, 2014.

[123] Y. Liu, Z. Su, X. Luan, D. Song, and L. Han, "Motion analysis and fuzzy-PID control algorithm designing for the pitch angle of an underwater glider," *Int. J. Math. Comput. Sci.*, vol. 17, pp. 133–147, 2017.

[124] J. Cao, J. Cao, Z. Zeng, and L. Lian, "Nonlinear multiple-input-multiple-output adaptive backstepping control of underwater glider systems," *Int. J. Adv. Robot. Syst.*, vol. 13, no. 6, pp. 1–14, 2016.

[125] D. Song, T. Guo, H. Wang, Z. Cui, and L. Zhou, "Pitch angle active disturbance rejection control with model compensation for underwater glider," in *Proc. Int. Conf. Intell. Robot. Appl.*, Wuhan, China, 2017, pp. 745–756.

[126] Z. Su, M. Zhou, F. Han, Y. Zhu, D. Song, and T. Guo, "Attitude control of underwater glider combined reinforcement learning with active disturbance rejection control," *J. Mar. Sci. Tech.*, vol. 24, no. 3, pp. 686–704, 2019.

[127] T. Guo, D. Song, K. Li, C. Li, and H. Yang, "Pitch angle control with model compensation based on active disturbance rejection controller for underwater gliders," *J. Coast. Res.*, vol. 36, no. 2, pp. 424–433, 2020.

[128] N. Mahmoudian and C. Woolsey, "Underwater glider motion control," in *Proc. IEEE Conf. Decis. Control*, Cancun, Mexico, 2008, pp. 552–557.

[129] F. Zhang and X. Tan, "Passivity-based stabilization of underwater gliders with a control surface," *J. Dyn. Syst. Meas. Control*, vol. 137, p. 6, 2015.

[130] S. Yan, R. Zhang, S. Yang, W. Niu, Y. Zhang, and B. Li, "Modelling and control optimization for underwater glider of variable buoyancy processed in vertical plane," *China Mech. Eng.*, vol. 33, no. 1, pp. 109–117, 2022.

[131] M. G. Joo and Z. Qu, "An autonomous underwater vehicle as an underwater glider and its depth control," *Int. J. Control Autom.*, vol. 13, no. 5, pp. 1212–1220, 2015.

[132] B. Claus and R. Bachmayer, "Energy optimal depth control for long range underwater vehicles with applications to a hybrid underwater glider," *Auton. Robot.*, vol. 40, no. 7, pp. 1307–1320, 2016.

[133] S. Guo, T. Fukuda, and K. Asaka, "A new type of fish-like underwater micro-robot," *IEEE/ASME Trans. Mechatronics*, vol. 8, no. 1, pp. 136–141, 2003.

[134] A. Keow and Z. Chen, "Modeling and control of artificial swimming bladder enabled by IPMC water electrolysis," GA, USA pp. V001T04A009, 2018.

[135] T. I. Um, Z. Chen, and H. Bart-Smith, "A novel electroactive polymer buoyancy control device for bio-inspired underwater vehicles," in *Proc. IEEE Int. Conf. Robot. Autom.*, 2011, pp. 172–177.

[136] M. Makrodimitris, I. Aliprantis, and E. Papadopoulos, "Design and implementation of a low cost, pump-based, depth control of a small robotic fish," in *Proc. IEEE/RSJ Int. Conf. Intell. Robots Syst.*, IL, USA, 2014, pp. 1127–1132.

[137] N. Mahmoudian, J. Geisbert, and C. Woolsey, "Approximate analytical turning conditions for underwater gliders: Implications for motion control and path planning," *IEEE J. Ocean Eng.*, vol. 35, no. 1, pp. 131–143, 2010, doi: 10.1109/JOE.2009.2039655.

[138] A. Caiti, V. Calabro, S. Grammatico, A. Munafò, and S. Geluardi, "Switching control of an underwater glider with independently controllable wings," *IFAC Proc.*, vol. 45, no. 27, pp. 194–199, 2012.

[139] H. Yang and J. Ma, "Sliding mode tracking control of an autonomous underwater glider," in *Proc. IEEE Int. Conf. Comput. Appl. Syst. Model*, Taiyuan, China, 2010, pp. 555–558.

[140] I. Abraham and J. Yi, "Model predictive control of buoyancy propelled autonomous underwater glider," in *Proc. Am. Control Conf.*, IL, USA, 2015, pp. 1181–1186.

[141] A. Caiti, V. Calabro, S. Geluardi, S. Grammatico, and A. Munafo, "Switching control of an underwater glider with independently controllable wings," *Inst. Mech. Eng. Part M. J. Eng. Marit. Envir.*, vol. 28, no. 2, pp. 136–145, 2014.

[142] Y. Zhang, L. Zhang, and T. Zhao, "Discrete decentralized supervisory control for underwater glider," in *Proc. Int. Conf. Intelligent Syst. Design Appl.*, China, vol. 2, 2006, pp. 103–106.

[143] Y. Zhang, J. Tian, D. Su, and S. Wang, "Research on the hierarchical supervisory control of underwater glider," in *Proc. IEEE/RSJ Int. Conf. Intell. Robots Syst.*, Beijing, China, 2006, pp. 5509–5513.

[144] K. Isa and M. R. Arshad, "Neural network control of buoyancy-driven autonomous underwater glider," *Recent Adv. Robot. Autom.*, pp. 15–35, 2013.

[145] N. Tan, Z. Zhong, P. Yu, Z. Li, and F. Ni, "A Discrete Model-Free Scheme for Fault Tolerant Tracking Control of Redundant Manipulators," *IEEE Trans. Ind. Inform.*, pp. 1–1, 2022, doi: 10.1109/TII.2022.3149919.

[146] Z. Li and S. Li, "An L_1-norm based optimization method for sparse redundancy resolution of robotic manipulators," *IEEE Trans. Circuits Syst. II Express Briefs*, vol. 69, no. 2, pp. 469–473, 2022, doi: 10.1109/TCSII.2021.3088942.

[147] J. Yu, F. Zhang, A. Zhang, W. Jin, and Y. Tian, "Motion parameter optimization and sensor scheduling for the sea-wing underwater glider," *IEEE J. Ocean Eng.*, vol. 38, no. 2, pp. 243–254, 2013.

[148] D. Xue, Z. Wu, Y. Wang, and S. Wang, "Coordinate control, motion optimization and sea experiment of a fleet of Petrel-II gliders," *Chin. J. Mech. Eng.*, vol. 31, no. 1, pp. 1–15, 2018.

[149] M. Yang, Y. Wang, S. Wang, S. Yang, Y. Song, and L. Zhang, "Motion parameter optimization for gliding strategy analysis of underwater gliders," *Ocean Eng.*, vol. 191, 2019.

[150] Y. Song, X. Xie, Y. Wang, S. Yang, W. Ma, and P. Wang, "Energy consumption prediction method based on LSSVM-PSO model for autonomous underwater gliders," *Ocean Eng.*, vol. 230, 2021.

[151] X. Fu, L. Lei, G. Yang, and B. Li, "Multi-objective shape optimization of autonomous underwater glider based on fast elitist non-dominated sorting genetic algorithm," *Ocean Eng.*, vol. 157, pp. 339–349, 2018.

[152] C. Li, P. Wang, T. Li, and H. Dong, "Performance study of a simplified shape optimization strategy for blended-wing-body underwater gliders," *Int. J. Nav. Arch. Ocean Eng.*, vol. 12, pp. 455–467, 2020.

[153] H. Dong, P. Wang, and Y. Zhang, "Discrete optimization design for cabin-skeleton coupling structure of blended-wing-body underwater glider," *China J. Ship Res.*, vol. 16, no. 4, pp. 70–78, 2021.

[154] N. Zhang, P. Wang, and B. Song, "Shape optimization for blended-wing-body underwater glider using improved Kriging-HDMR," *J Unmanned Undersea Syst.*, vol. 27, no. 5, pp. 496–502, 2019.

[155] M. Yang, Y. Wang, S. Yang, L. Zhang, and J. Deng, "Shape optimization of underwater glider based on approximate model technology," *Appl. Ocean Res.*, vol. 110, 2021.

[156] J. Cao, J. Cao, Z. Zeng, and L. Lian, "Optimal path planning of underwater glider in 3D Dubins motion with minimal energy consumption," in *Proc. IEEE OCEANS*, Shanghai, China, 2016, pp. 1–7.

[157] Y. Liu, J. Ma, N. Ma, and G. Zhang, "Path planning for underwater glider under control constraint," *Adv. Mech. Eng.*, vol. 9, no. 8, pp. 1–9, 2017.

[158] M. Arima, N. Ichihashi, and Y. Miwa, "Modelling and motion simulation of an underwater glider with independently controllable main wings," in *Proc. IEEE OCEANS*, Bremen, Germany, 2009, pp. 1–6.

[159] Y. Li, D. Pan, Q. Zhao, Z. Ma, and X. Wang, "Hydrodynamic performance of an autonomous underwater glider with a pair of bioinspired hydro wings–a numerical investigation," *Ocean Eng.*, vol. 163, pp. 51–57, 2018.

[160] A. T. Khan, S. Li, and X. Cao, "Control framework for cooperative robots in smart home using bio-inspired neural network," *Measurement*, vol. 167, p. 108253, Jan. 2021.

[161] A. T. Khan, S. Li, and X. Cao, "Human guided cooperative robotic agents in smart home using beetle antennae search," *Sci. China Inf. Sci.*, vol. 65, no. 2, p. 122204, Jan. 2022.

[162] A. T. Khan, S. Li, and X. Zhou, "Trajectory Optimization of 5-Link Biped Robot Using Beetle Antennae Search," *IEEE Trans. Circuits Syst. II Express Briefs*, vol. 68, no. 10, pp. 3276–3280, 2021.

[163] Z. Li, S. Li, and X. Luo, "An overview of calibration technology of industrial robots," *IEEE/CAA J. Autom. Sin.*, vol. 8, no. 1, pp. 23–36, Jan. 2021.

[164] X. Luo, W. Qin, A. Dong, K. Sedraoui, and M. Zhou, "Efficient and high-quality recommendations via momentum-incorporated parallel stochastic gradient descent-based learning," *IEEE/CAA J. Autom. Sin.*, vol. 8, no. 2, pp. 402–411, Feb. 2021.

[165] X. Luo, Z. Wang, and M. Shang, "An instance-frequency-weighted regularization scheme for non-negative latent factor analysis on high-dimensional and sparse data," *IEEE Trans. Syst. Man Cybern. Syst.*, vol. 51, no. 6, pp. 3522–3532, Jun. 2021.

[166] C. C. Wall, C. Lembke, and D. A. Mann, "Shelf-scale mapping of sound production by fishes in the eastern Gulf of Mexico, using autonomous glider technology," *Mar. Ecol. Prog. Ser.*, vol. 449, pp. 55–64, Mar. 2012.

[167] L. K. P. Schultze, L. M. Merckelbach, and J. R. Carpenter, "Turbulence and mixing in a shallow shelf sea from underwater gliders," *J. Geophys. Res. Oceans*, vol. 122, no. 11, pp. 9092–9109, 2017.

CHAPTER 2

Design and Implementation of Typical Gliding Robotic Dolphins

2.1 INTRODUCTION

Dolphins are gifted swimmers with outstanding aquatic locomotion capabilities [168]–[170]. They can effortlessly swim fast, maneuver flexibly, and even complete astonishing acrobatics; for example, they could easily obtain a high speed over 11 m/s and then leap up to several meters and perform some elegant acrobatic stunts [171]. These attractive skills are endowed by their innate streamlined bodies, skeletal anatomies, and propulsive mechanisms, which are products of a long period of biological evolution. Considering such surprising swimming skills of the dolphin, researchers have focused on dolphin-like swimming for several years to explore a practical, effective, and flexible propulsive mechanism. They have been trying to realize dolphin-like swimming performance via various researches of robotic dolphins [172]–[176], from the perspective of propulsive mechanism design, locomotion modeling, and control algorithm. For example, Nakashima *et al.* [174] implemented a robotic dolphin with rotating flippers and dorsal fin and achieved quasi-three-dimensional (quasi-3D) maneuverability. Ren *et al.* proposed an average propulsive

DOI: 10.1201/9781003347439-2

speed implementation method for a bio-inspired robotic dolphin [175]. Yu *et al.* [176] realized a fast swimming speed (2.9 BL/s or 2.07 m/s, BL for body length) and replicated acrobatic-like dolphin leaping behaviors based on an integrative control method. With the improvement of mechatronic systems and control algorithms, speed, and maneuverability of robotic dolphins also evolve. Therefore, owing to their excellent swimming performance and distinctive characteristics, robotic dolphins have potential for many real-world applications, such as exploration and rescue in complex underwater environments. However, their short range and limited endurance are potential obstacles lying in the path of practical applications.

By contrast, underwater gliders are well known for their characteristics of long range and strong endurance [177]–[178]. Their extraordinary energy-efficiency, in essence, originates from the buoyancy-driven mechanism: propel by the synthetic effect of net buoyancy and hydrodynamic forces and move in a sawtooth-like path. Energy is consumed only when they adjust buoyancy to switch between diving and surfacing. Therefore, the underwater glider could glide for a long distance with limited energy. Nevertheless, as a sword has double blades, both the speed and maneuverability of gliders are much lower. Therefore, how to create a gliding robotic dolphin with hybrid motions for deepsea applications and how the controllable fins enrich the gliding maneuvers are worthy of intensive investigation.

The main purpose of this chapter is to offer an innovative concept for the gliding underwater robots, namely, gliding robotic dolphin, to implement both high maneuverability and long endurance. As a combination of robotic dolphins and underwater gliders, the gliding robotic dolphin could not only realize fast and flexible dolphin-like swimming but also glide for a long distance due to the buoyancy-driven system. The main contributions are as follows:

- We present the system development of four newly designed gliding robotic dolphins, including their overview and detailed electromechanical scheme, offering a novel idea for the future application of gliding underwater robots.

- Extensive computational fluid dynamics (CFD) simulations are conducted, which not only can assist in optimizing the shape design to improve gliding efficiency but also can obtain the key hydrodynamic coefficients to support the theoretical research.

- Based on the designed prototypes, extensive aquatic experiments and analysis are carried out to show the multimodal motion performances, including the gliding and dolphin-like modes.

The remainder of this chapter is organized as follows. Section 2.2 provides the mechanical design of typical gliding robotic dolphins in detail. Section 2.3 presents the CFD simulation and analysis of the gliding robotic dolphin. Section 2.4 shows the experimental results and discussion of the motion performances. Section 2.5 presents our conclusions.

2.2 SYSTEM DEVELOPMENT OF TYPICAL GLIDING ROBOTIC DOLPHINS

For the applications in lakes, rivers, reservoirs, and other waters, we aim to design a modular and high-safety gliding underwater robot with different lengths and weights, exploring the new autonomous underwater vehicle (AUV) that combines high maneuverability and long endurance, and accomplishing the large-scale autonomous cruise and ocean applications.

The gliding robotic dolphin takes the Orca (refer to Figure 2.1) as the prototype for its appearance design. According to biological data, killer whales belong to the genus *Orcinus* of the Delphinidae, and are the largest species of dolphin. Adult male killer whales can grow to a body length of more than 9 m and a weight of 10 tons. Compared with other common dolphins such as bottlenose dolphins, killer whales have a slightly rounded head and a spindle-shaped body. On one hand, the shape of the killer whale is selected in this scheme due to convenient machining. On the other hand, the round head is beneficial to improve the utilization of internal space and facilitate the installation of front equipment. Besides, it

FIGURE 2.1   Natural killer whale.

should be noted that the gliding robotic dolphin designed in this scheme reduces the proportion of male killer whales. The flippers, flukes, dorsal fin, and other fin surfaces adopt national advisory committee for aeronautics (NACA) series low-speed airfoil NACA-0018, with a relative thickness of 18%. For specific data, please refer to [179].

Compared with traditional underwater gliders and biomimetic ones [62], the gliding robotic dolphin has several special characteristics as follows:

1. *Flapping posterior body*: By means of the dolphin-like swimming mechanism, the gliding robotic dolphin could easily perform a much faster and more flexible motion than traditional underwater gliders.

2. *Controllable flippers*: Traditional underwater gliders with fixed wings often utilize an internal moveable mass to adjust their attitude in gliding motion that may bring a slow response and require a sufficient large space. Instead, the robotic dolphin adopts controllable flippers to adjust the gliding attitude. Moreover, the flippers can also provide enough thrust for multimodal locomotion such as forward swimming, turning, descending, and ascending.

3. *Flatten and horizontal fluke*: It could not only produce a high propulsive speed in dolphin-like motion but also provide considerable pitching torques due to the relatively larger force arm.

These fascinating features make the gliding robotic dolphin can perform three kinds of motions including gliding motion, dolphin-like swimming with two-joint flapping, and stable flipper propulsive motion. To the best of our knowledge, this is the first time that three hybrid motions have been successfully performed on the same robotic platform. Furthermore, the robot could adaptively choose the appropriate motion mode relying on the mission and underwater environments.

2.2.1 A Miniature Dolphin-Like Underwater Glider

In this section, we introduce a simple testify dolphin-like underwater glider for gliding motion. As we have confidence and experience in the mechanical design and motion control for a robotic dolphin, the prototype in this section is only developed to realize gliding motion without dolphin-like dorsoventral joints. Besides, to testify how the flippers and flattened fluke affect the gliding motion, including gliding attitude, gliding speed, and the like, controllable flippers and flattened fluke are also designed for the robot. Note that traditional internal moveable masses are specially removed to highlight that the dolphin-like glider could obtain enough pitching torques from the internal buoyancy-driven system, controllable flippers, and flattened fluke.

The mechanical design of the dolphin-like underwater glider is schematically shown in Figure 2.2. Generally, the dolphin-like underwater glider is 0.37 m in length and weighs 0.75 kg. A rigid well-streamlined body modeled after killer whale is adopted for a better space utilization rate and lift–drag ratio. The translucent body is made of acrylonitrile butadiene styrene copolymers and could house control circuits, battery packs,

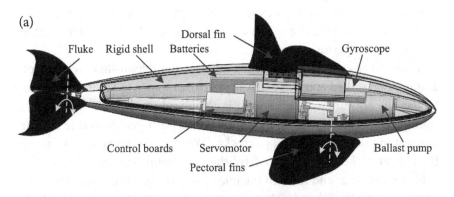

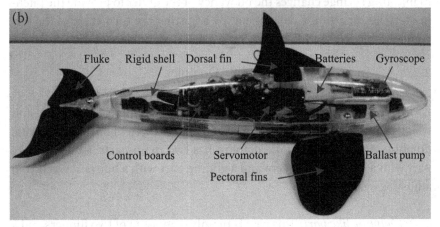

FIGURE 2.2 Mechanical design of the miniature dolphin-like glider. (a) Conceptual design. (b) Robotic prototype.

Ballast pump Rubber Fixed bearing Push rod Rocker Servomotor

FIGURE 2.3 Mechanical design of the buoyancy-driven system.

gyroscope sensors, communicating modules, buoyancy-driven system, and so on. The flippers and flattened fluke, made of polypropylene, can manually be regulated to an expected turn angle via a screw arbor structure. Besides, these fins are a low-speed airfoil, NACA0018, for a better hydrodynamic performance.

To obtain an expected net buoyancy change, the dolphin-like glider adopts a simple and skilled buoyancy-driven system, as shown in Figure 2.3. In particular, the buoyancy-driven system is composed of an injecting syringe, a waterproof rubber, a digital servomotor, and a push–pull structure with an aluminum push rod and a copper rocker. When the servomotor is working, the copper rocker is turned to drive the aluminum push rod to make the rubber move back and forth in the injecting syringe. Meanwhile, the water in injecting syringe changes the buoyancy of the glider to provide the pitching torques for attitude adjustment. By adjusting the turn angle of the servomotor, the dolphin-like glider could gain an accurate buoyancy change.

2.2.2 A 1-m-Scale Gliding Robotic Dolphins

2.2.2.1 Type I

As illustrated in Figure 2.4, the gliding robotic dolphin is 0.83 m long and weighs 8.86 kg. As a typical combination of underwater gliders and robotic dolphins, the gliding robotic dolphin has two main parts: a dolphin-like part with a propulsive system and a gliding part with a buoyancy-centroid adjusting system.

1. *Dolphin-like part*: This part is mainly composed of two flippers and a tail cabin. Two flippers can be used to perform the deflecting motion and median and/or paired fin (MPF) motion. The robot can use this

motion to achieve not only forward/backward movements but also a small-radius turn via flapping the two flippers in specific ways. Besides, the body and/or caudal fin (BCF) motion can be implemented by flapping the waist and caudal joints. The corresponding motor selections of dolphin-like parts are shown in Figure 2.4(a).

2. *The gliding part*: This part of an injector-based buoyancy-driven mechanism and a slider-based pitching mechanism. The detail mechanical structure is figured in Figure 2.4(a). In order to fulfill the suction and

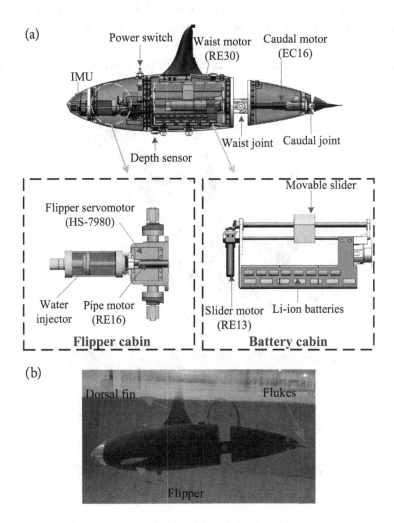

FIGURE 2.4 Overview of the gliding robotic dolphin I. (a) Conceptual design. (b) Robotic prototype.

drainage, a DC brush motor (RE16) is used to drive the gear for further moving the piston. Besides, a rotating plug is employed to guarantee waterproof. Regarding the moving slider, its basic principle is almost the same as the injector's, and the difference lies in the lower power motor (RE13). Hence, the injector and slider can be controlled via the position mode of motors, which can realize the relatively precise buoyancy control via calculating the volume change.

2.2.2.2 Type II

For simplified mechanical design and easy installation, the gliding robotic dolphin adopts a distinctive modularity concept. As shown in Figure 2.5, the gliding robotic dolphin consists of six principal cabinets according to their own function: head cabinet, flipper cabinet, mission cabinet,

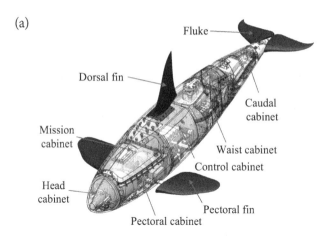

FIGURE 2.5 Overview of the gliding robotic dolphin II. (a) Conceptual design. (b) Robotic prototype.

control cabinet, waist cabinet, and caudal cabinet. These cabinets adopt wired electrical connections and wireless command connections, which can effectively improve the waterproof capability. Generally, the gliding robotic dolphin is 1.13 m long and weighs 18.20 kg. Meanwhile, a low-speed airfoil, NACA0018, is also employed to design the flippers and flattened fluke for a better hydrodynamic performance.

On one hand, to realize excellent dolphin-like swimming, an effective and powerful propulsive unit is designed. The propulsive unit employs two pitching joints respectively called waist and cauda to execute symmetrical sinusoidal dorsoventral oscillations for enough propulsive thrusts. Two powerful DC motors (Maxon EC-4pole) with a fast turn speed and strong torque are utilized to drive the posterior posterior-body up-and-down locomotion. Besides, the robotic dolphin also employs a yawing joint in waist cabinet to achieve planar turning maneuvers. Alternatively, the turning maneuvers can also be realized through the flippers in a high propulsive speed. In order to change the turning direction from roll to pitch, three groups of reducing bevel gear sets are employed by the body joints. These bevel gear sets are all made of titanium alloy, which can not only reduce the weight but also avoid getting rusty. Notice that the adopted DC motors should be powered off to save energy when they need not work such as in a steady gliding motion. In this case, a brake system is utilized to keep the body locked.

On the other hand, a practical buoyancy-driven system that changes the buoyancy through draining off and pumping the water is developed to realize the gliding motion. As shown in Figure 2.6, the buoyancy-driven

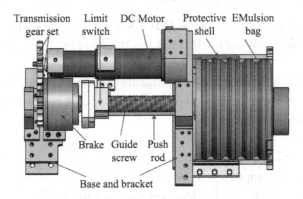

FIGURE 2.6 Mechanical design of the buoyancy-driven system.

system consists of a DC motor, transmission gear set, a guide screw with a brake, a push rod, and an emulsion bag in a protective shell. When changing the buoyancy of the gliding robotic dolphin, the DC motor turns to drive the guide screw to move back and forth through the transmission gear set. Simultaneously, the push rod begins to push and pull the emulsion bag into the protective shell. With the moving of the emulsion bag, the water is pumped and drained off to change the buoyancy of the robotic dolphin. Specially, a limit switch fixed on the top of the guide screw is utilized to keep the running range of the guide screw. Similar to the propulsive unit, the buoyancy-driven system is only powered on when needing to work. In order to gain additional pitching torques, the buoyancy-driven system is fixed in the head of the flipper cabinet. Therefore, the changing volume of the emulsion bag could provide pitching torques for attitude adjustment. Thus, an internal moveable mass which is necessary for traditional underwater gliders can be removed, and the robotic dolphin can effectively regulate its attitude with the pitching torques from the buoyancy-driven system and the controllable flippers and horizontal fluke.

2.2.3 A 1.5-m Gliding Robotic Dolphin with 3 MPa pressure

Aiming for deep-sea applications, we develop a gliding robotic dolphin with higher performance requirements, and its design goals are as follows:

1. The maximum diving depth can reach 300 m, the body length and the total mass are about 1.5 m and 55 kg, respectively.

2. The maximum gliding speed is not less than 0.3 m/s (i.e., 0.2 BL/s), and the maximum travel speed is not less than 1.5 m/s (i.e., 1 BL/s).

3. The gliding range is not less than 1500 km, and the swimming range under dolphin-like mode is not less than 40 km.

In view of the stated design goals, this section introduces the overall design scheme from the aspects of shape layout, joint configuration, material selection, and sealing method.

2.2.3.1 Overall Shape

According to the approximate body ratio of the killer whale, the appearance and dimensions of the designed gliding robotic dolphin are shown in Figure 2.7. The shape and relative size of the flippers, flukes, and dorsal

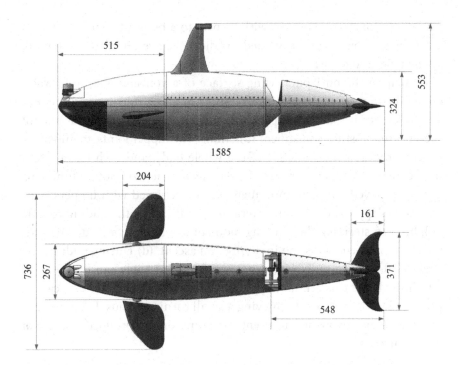

FIGURE 2.7 The appearance and dimensions of the designed gliding robotic dolphin.

fins are similar to those of killer whales. The vertical cross-sectional shape of these fins refers to the NACA-0018 airfoil with a relative thickness of 18%.

2.2.3.2 Joint Configuration

In order to enable the gliding robotic dolphin to swim quickly through a propulsive system, two joints, namely, the waist and tail joints, are set in the back half of the body. Concretely, the waist joint is roughly located at 2/3 of the entire body and can drive the tail to swing up and down. The tail joint is located at the end of the body for rotating the flukes. Theoretically, the more tail joints, the closer the dorsal-abdominal movement is to the porpoise wave. However, this design needs more space to place the driven device, and the electromechanical design will be complicated. The functions of the tail joint mainly include the following: (a) Cooperating with the waist joint to produce propulsive movement, it can generate sufficient forward propulsion by flapping up and down. (b) The tail joint has a relatively long moment arm with respect to a buoyant center. When in a gliding motion, it can be applied to adjust the overall lift-to-drag ratio, pitch angle, and glide angle.

Similar to the horizontal wing surface of a conventional underwater glider, the flippers of the gliding dolphin can play a role in providing lift. The difference is that the flippers are movable so that they own abundant exercise and posture adjustment functions. The flipper joints are independently located at the front 1/4 of the whole body, and each has a pitching degree of freedom. Through the setting, a variety of motion functions can be achieved: (a) MPF movement can be achieved with flippers' synchronous flapping. (b) Via unilateral flipper flapping, the gliding robotic dolphin can steer. (c) The pitching moment can be generated during propulsive movement, assisting in diving and ascent. (d) During gliding, the overall lift-to-drag ratio, pitch angle, and gliding angle can be adjusted. (e) Through the differential flapping of flippers, the yaw moment can be produced to steer, further achieving a small turning radius. It can be seen that the previously mentioned joint configuration can produce a wealth of motion modes.

2.2.3.3 Material Selection

Given the overall geometric shape of the gliding robotic dolphin, its displacement and buoyancy are basically determined. When designing, it is necessary to ensure that the gravity is less than the buoyancy, leaving a load margin for the internal load and counterweight. Therefore, light materials with low density should be preferred when selecting materials. Considering that the diving depth of the gliding robotic dolphin needs to reach 300 m, that is, it needs to be able to withstand the external water pressure of 3 MPa, the shell material should have strong pressure resistance. In order to minimize its own weight and ensure the shell strength, we choose 5A06 hard aluminum alloy as the shell material, with a density of 2.7×10^3 kg/m^3. Besides, its tensile strength and conditional yield strength are higher than 315 MPa and 160 MPa, respectively. In order to avoid deformation of the shell during the machining process, the shell thickness is set to 6 mm after pressure calculation and instability analysis. Moreover, we further carry out anodizing treatment on the shell to improve the corrosion resistance and wear resistance of the shell surface.

The materials of the parts such as the fasteners and support frames inside the robot are made of aviation hard aluminum LY12, and its tensile strength can reach 425 MPa. For components with greater bearing capacity, such as transmission gear sets and shaft seats, they are made of TC4 titanium alloy with a density of 4.5×10^3 kg/m³ and a tensile strength of 895 MPa. It has advantages of high strength, low density, good toughness, and corrosion resistance. Furthermore, the transmission parts exposed to water are made of titanium alloy and stainless steel with hard chromium plating to prevent the corrosion of seawater and oxygen. The fin surface is made of polypropylene (polypropylene, density 0.91×10^3 kg/m³), which has a density close to that of water. The oil bladder is chosen with flexible TPU (thermoplastic urethane) material, which has excellent properties such as toughness, oil resistance, and aging resistance. It is capable of working conditions of the oil bladder: frequent expansion and contraction and high pressure. The head fairing used to protect the external oil bladder is 3D-printed with acrylonitrile butadiene styrene (ABS) plastic to save material and facilitate processing. In addition, polyformaldehyde is selected as the processing material of the conformal decorative parts or parts protection cover. This material has the characteristics of toughness and dimensional stability and can avoid the problem of difficult installation due to deformation. The main materials used in the mechanical structure of the gliding robot dolphin are shown in Table 2.1.

2.2.3.4 Sealing Method

The reliability and safety of the gliding robotic dolphin are closely related to the sealing method. Water leakage during debugging or experimentation is devastating to the electronic components inside the robot, while

TABLE 2.1 Mechanical Materials and Their Properties

Material name	Characteristic	Application object
5A06 aluminum	Small density, high strength	Shell
LY12	Small density, high strength	Small bearing parts
TC4	Low density, high strength, anti-rust	Heavy-duty components
Stainless steel	High strength, anti-rust	Parts with large bearing capacity such as shafts and shaft seats
Polypropylene	Similar in density to water	Flippers, dorsal, and flukes
TPU	High tension, oil resistance, aging resistance	Oil bladder
ABS plastic	Thermoplastic, impact resistant	Head fairing
Polyoxymethylene	Tough, dimensionally stable, smooth	Conformal decorations, protective covers

water leakage in actual use may result in the loss of the entire gliding robotic dolphin. Since the gliding robotic dolphin needs to withstand up to 3 MPa water pressure, and there are four joints that need to be dynamically sealed, the sealing method has become one of the most challenging problems in the entire design.

According to whether the sealing object is relatively static or moving, it can be divided into static sealing and dynamic sealing. Generally, static seals are made of fluorine rubber O-rings with a tensile strength of 10 MPa for the joints of the shells, and it is characterized by high-pressure resistance, corrosion resistance, aging resistance, and a wide operating temperature range. The static seal adopts the end face seal design to improve the reliability of the seal, further facilitating the mechanical assembly.

The difficulty of seal design lies in the dynamic seal at the drive shaft. In this section, a rotating pan plug is particularly applied for dynamic sealing, whose dynamic and static load can reach 15 MPa and 25 MPa, respectively. The rotating pan plug is a U-shaped envelope with a spring inside, which can seal a rotating or reciprocating shaft in one direction. The rotating pan plug is attached to the wall surface of the shaft, and the shell through the sealing lips on both sides of the U-shaped groove. On one hand, when the external water pressure is low, the outward tension of the sealing lip comes from the spring in the U-shaped groove. On the other hand, when the external water pressure is high, the tension can be provided by the water pressure, and the greater the water pressure, the tighter the sealing lip fits. Therefore, the rotating pan plug cleverly uses the external water pressure to dynamically compensate for the force of the fit, and it can play a sealing role regardless of the external water pressure or the dynamic and static of the shaft. In addition, the tension exerted by the spring and water pressure can overcome the slight unevenness of the shaft surface and the wear of the sealing lip. The rotating pan plug is made of polytetrafluoroethylene (PTFE), which has excellent wear resistance and corrosion resistance.

After many times of pressure test, a complete design of the dynamic seal at the rotating shaft can be given, as shown in Figure 2.8. More important, it is necessary to particularly emphasize the following points:

1. In order to prevent gaps caused by the shaking of the shaft and cause water leakage, the rotating shaft needs to be fixed by two or more bearings. Since the bearing is exposed to water, ceramic material should be used here to prevent rust and corrosion.

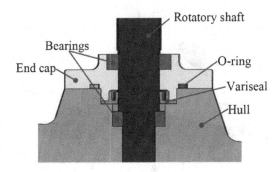

FIGURE 2.8   The illustration of the dynamic seal.

2. To prevent liquid from entering the gap between the end cover and the shell, an O-ring should be added between the end cover and the shell for static sealing.

3. The smoother the surface of the rotating shaft, the smaller the wear on the rotating pan plug. Hence, the sealing performance is more reliable. Generally, the surface roughness of the shaft must satisfy $Ra \leq 0.4$ μm. Note that when the external water pressure is high, the pressure at the joint between the sealing lip of the rotating pan plug and the shaft surface will be large. If the shaft surface is rough, the pan plug will be worn out quickly even though the PTFE material has the characteristic of being wear-resistant. In this section, the chrome-plated optical shaft is adopted as the external drive shaft, and the coating on the surface is quite smooth and hard.

2.2.3.5 Electromechanical System Design

In order to facilitate processing and assembly, the gliding robotic dolphin is divided into three cabins from front to back, named flipper cabin, control cabin, and tail cabin. In this section, we describe the mechanism design of each cabin in turn.

2.2.3.5.1 Flipper Cabin Design   The flipper cabin is located on the head of the gliding robotic dolphin, and its internal and external structures are shown in Figure 2.9. The cabin is composed of the top and bottom shells, which are sealed by O-rings. Two flippers are located on both sides of the cabin. The front of the top shell is equipped with a miniature forward-looking sonar (Model: Tritech Micron Sonar, detection range:

(a)

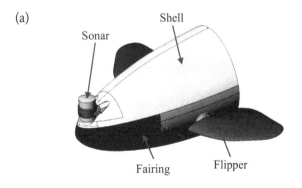

(b)

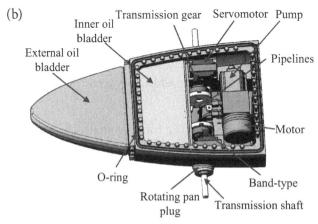

FIGURE 2.9 The illustration of the flipper cabin. (a) Exterior structure. (b) Interior structure.

0.3~75 m, scanning range: 360°) for obstacle avoidance and target tracking. Furthermore, a fairing is installed in the front of the bottom shell. On one hand, it can protect the outer oil bladder. On the other hand, the overall streamline and hydrodynamic characteristics can be maintained via it. Besides, on the fairing, there are some small holes communicating with the external liquid so that the outer oil bladder is exposed to water. The flippers cabin mainly accommodates two mechanisms: a buoyancy adjustment mechanism and a flipper-driven mechanism.

The buoyancy adjustment mechanism is the key mechanism for realizing gliding motion, which endows the robot with the capability of gliding. This mechanism can adjust the overall net buoyancy of the robot, further achieving floating and diving. Due to the effect of hydrodynamics, the gliding robotic dolphin does not float up and down vertically but glides at

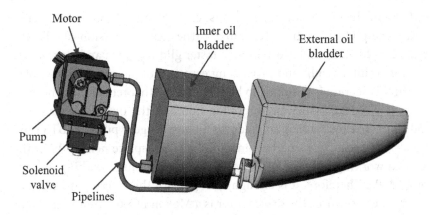

FIGURE 2.10   The illustration of the buoyancy-driven system.

a certain gliding angle. The buoyancy adjustment mechanism can change the volume of the outer oil bladder by oil circulation, thereby adjusting the overall buoyancy, and its composition structure is shown in Figure 2.10. Compared with the suction and drainage buoyancy adjustment mechanism in literature [62], the design in this section can make the cabin form an environment isolated from external liquids and can prevent the suction and drainage channels from being blocked by sediment or attached marine organisms.

Specifically, the mechanism is composed of two flexible oil bladders, a two-way hydraulic pump, a drive motor, a solenoid valve, an oil pipe, and a fixed part. Among them, the inner oil bladder is inside the cabin, while the outer oil bladder is exposed to water. The two oil bladders contain hydraulic oil, and the total oil volume is equal to the volume of a single oil bladder. The motor drives the hydraulic pump to rotate, and the oil can be transferred between two oil bladders. When the motor rotates forward, the hydraulic pump sucks oil, and the buoyancy decreases. When the overall buoyancy is less than gravity, the gliding robotic dolphin starts to dive. On the contrary, the hydraulic pump discharges oil as the motor reverses, leading to positive buoyancy. When the overall buoyancy is greater than gravity, the gliding dolphin starts to float. The solenoid valve is only opened when the buoyancy is adjusted, avoiding the external water pressure to squeeze oil into the inner oil bladder.

The capacity of the oil bladder determines the upper limit of net buoyancy and further determines the maximum gliding speed. In general, the maximum net mass of a typical underwater glider accounts for 0.5%

to 0.8% of the overall displacement (such as Slocum Electric and Spray), and some gliders reach more than 1% (the reduced version of STERNE reaches 1.26%). In order to achieve a faster gliding speed and leave a margin for counterweight and processing errors, the capacity of the oil bladder in this section is designed to be 1.925 L. The theoretical maximum net mass is 0.963 kg, which accounts for 1.66% of the overall displacement (58.1 kg). The hydraulic pump adopts a two-way gear pump (CBTS type), which has a relatively small volume (0.41 kg), a displacement of 0.25 mL/r, a rated working pressure of 21 MPa, and a maximum working pressure of 28 MPa. Therefore, it is enough to cope with the external maximum 3 MPa water pressure. The drive motor is a Maxon EC60 motor with a rated power of 100 W, a rated speed of 3850 rpm, and a rated torque of 0.221 N·m. The solenoid valve (Model: Wandfluh SIN29V) is normally closed to save power. The measured maximum speed of the motor during operation is 3200 r/min, and the maximum oil flow speed is 13.3 mL/s. Hence, it takes about 144 s to empty the oil bladder.

The flippers of the gliding robotic dolphin can produce multiple motion modes and can provide lift or be used for attitude adjustment during gliding. This feature is not available in the horizontal fixed wing on the conventional glider. The flipper compartment is equipped with two sets of independent-driven mechanisms, which are composed of a steering gear, a pair of mutually meshing cylindrical gears, a transmission shaft, and a holding brake. The structure is shown in Figure 2.11. The steering gear model is Hitec HS-7980TH with 4.4 N·m maximum torque, 352°/s maximum speed, and ±90° angle range. The flippers remain stationary during most time of the gliding motion, so a normally closed brake is added to the flipper-driven mechanism. When the flipper is not required to move, the

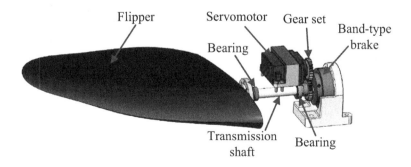

FIGURE 2.11   The illustration of the flipper mechanism.

brake can lock the drive shaft to avoid unnecessary energy loss caused by the torque output of the steering gear.

2.2.3.5.2 Control Cabin Design   The control cabin is located in the middle of the gliding robotic dolphin, connecting the front and the back. On one hand, it is responsible for the overall task execution, motion control, and data collection. On the other hand, it can provide power and command to the other cabins. In detail, the control cabin is equipped with a main control circuit, a communication module, a Global Position System (GPS) positioning module, a power management module, a lithium battery pack, and a pitch adjustment mechanism. The overall composition structure is shown in Figure 2.12. The dorsal fin fixed on the upper part of the shell is not movable. The main function of the dorsal fin of the natural killer

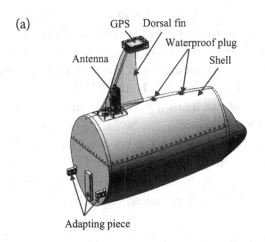

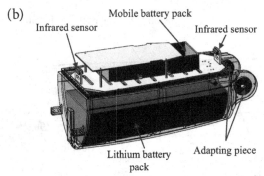

FIGURE 2.12   The illustration of the control cabin. (a) Exterior structure. (b) Interior structure.

whale is to provide an equilibrium sense, and the function of the dorsal fin here is reflected in three aspects: (a) Calm the posture during movement. (b) The cavity inside the dorsal fin can accommodate the communication antenna (wireless radio frequency), avoiding the communication signal being shielded by the metal cabin shell. At the same time, the dorsal fin can better maintain the overall bionic streamline. (c) The dorsal fin can be used to install the receiving antenna of GPS positioning module signal. Moreover, the top shell is equipped with a waterproof electrical connector, and the front and rear cabins are connected through a waterproof cable for power supply and communication. Furthermore, a high-energy density lithium battery pack is installed in this cabin, and it is composed of multiple lithium batteries (Model: Panasonic NCR18650B, single-cell voltage: 3.7 V, single-cell capacity: 3400 mAh) in series and parallel. The total mass, rated output voltage, and total capacity of the battery pack are 18 kg, 33.3 V, and 88 Ah, respectively. In order to enhance the rolling stability, the main part of the battery pack is placed on the bottom of the cabin. In addition, sensors such as the main control circuit board, communication module, power management module, and inertial navigation are all fixed on the circuit support board shown in Figure 2.12(b).

The pitch adjustment mechanism is the only movable mechanism in the control cabin. On one hand, it can be used alone to adjust the pitch angle of the gliding robotic dolphin. On the other hand, with aid of flippers and flukes, it can achieve decoupling control of the pitch and gliding angles. The structure is shown in Figure 2.13. The mechanism is composed of a mobile battery block, a drive motor, two infrared sensors, a linear

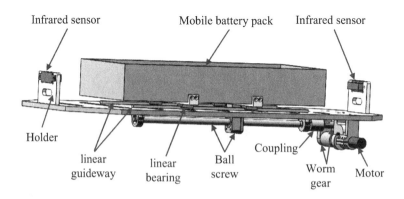

FIGURE 2.13 The illustration of movable mass.

guide, a linear bearing, a worm gear, a ball screw, and other transmission mechanisms. Specifically, the ball screw (Model: TBI SFK0802) can convert the rotation of the motor into the rectilinear motion of the mobile battery block, which makes the barycenter of the robot longitudinally shift, thereby changing the pitch angle. Linear guides and bearings can reduce friction, and further enable the battery block to move smoothly. Besides, worm gears (reduction ratio 4:1) play two roles: (a) Transmit rotation when the drive shaft directions are perpendicular to each other, reduce the output speed of the motor, and increase the output torque. (b) Thanks to its self-locking feature, the battery block can be kept in a stationary state without the need for a motor to maintain torque output, which can save energy. Theoretically, the position of the battery block can be calculated indirectly through the motor's movement. However, the mechanism gap is practically difficult to avoid, making it hard to obtain an accurate initial position of the battery block in this way. Therefore, an infrared sensor (Model: SHARP GP2Y0A41SK0F) is applied in the mechanism to directly measure the position, and its measurement range is 4~30 cm. Note that such sensors have a certain blind zone (here 4 cm). In order to eliminate the blind zone, we use two infrared sensors arranged at both ends of the mobile battery block. When both sensors detect the battery block, the data will be fused, which can reduce measurement errors and improve reliability. If only a single sensor detects the target, only its data are used as the measurement result.

The mass and maximum stroke of the battery block are 2.86 kg and 14 cm, respectively, so that the range of pitch angle adjustment is about ±15°. The selection and calculation process of the drive motor is as follows. Under normal circumstances, the range of pitch angle is within ±30°. Here, considering the extreme case, that is, when the pitch angle is ±90°, the driven torque required by the ball screw is

$$\tau_m = \frac{m_m \cdot I}{2\pi\mu} = 0.01 \, \text{N} \cdot \text{m}, \qquad (2.1)$$

where m_m denotes the mass of movable slider, and screw lead is $I = 2 \times 10^{-3}$ m. Besides, $\mu = 0.9$, $g = 9.8 \, \text{m/s}$.

2.2.3.5.3 Tail Cabin Design The tail joint compartment is located at the end of the gliding robotic dolphin, and mainly accommodates the driven mechanism of the waist and tail joints. The structure is shown in Figure 2.14. Both driven mechanisms are composed of drive motors,

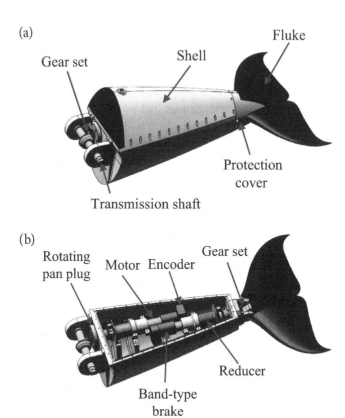

FIGURE 2.14   The illustration of the tail cabin. (a) Exterior structure. (b) Interior structure.

reducers, encoders, brakes, zero/limit photoelectric switches, and transmission mechanisms, as shown in Figure 2.15 (take the waist joint–driven mechanism as an example). The two motors are longitudinally fixed on the bottom shell, and the rotation is transmitted through bevel gears. The rotating shafts extending out of the shell are sealed by the rotating pan plug. According to the control experience of the robotic dolphin, the rotation ranges of the waist and tail joints are designed to be ±32° and ±60°, respectively. The configuration information of driven mechanisms is listed in Table 2.2. In order to indicate the zero position and limit position of the joint to the motor control system, a photoelectric switch (Model: GK102) is set in the corresponding position, and the pulse signal of the photoelectric switch is triggered by the opaque shutter that rotates with the motor output shaft.

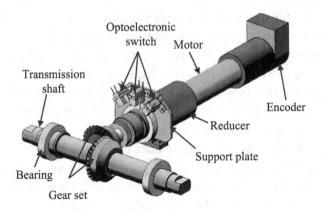

FIGURE 2.15 The illustration of the propulsive system.

TABLE 2.2 Driven Configuration Information for Waist and Tail Joints

Item	Waist Joint	Tail Joint
Motor type	Maxon EC-4Pole brushless motor	Maxon EC-4Pole brushless motor
Driver type	Copley ACM-090-24	Copley ACK-055-10
Motor power	120 W	90 W
Rated motor speed	16800 rpm	14900 rpm
Rated motor torque	0.054 Nm	0.044 Nm
Reducer reduction ratio	51:1	66:1
Bevel gear tooth ratio	22:30	20:30
Number of encoder lines	500	500

2.3 CFD SIMULATION AND ANALYSIS

In this section, we give a detailed CFD simulation and analysis to explore the gliding performance of the gliding robotic dolphin and obtain important hydrodynamic force coefficients.

Generally, hydrodynamic force coefficients such as lift coefficient, drag coefficient, and moment coefficient, could be determined using a variety of methods including airfoil theory, CFD methods, and flight tests. Here, CFD methods are employed to analyze the gliding performance of the robotic dolphin. Unlike the fixed wings of traditional underwater gliders, the flippers and flattened flukes of the gliding robotic dolphin can be manually controlled for expected pitching torques. Hence, we need separately compute hydrodynamic coefficients of lifts, drags, and pitching moments from the dolphin body, flippers, and horizontal flukes, which could be applied to the hydrodynamic analysis in gliding motion. For an accurate and convenient CFD simulation results, the commercial software ANSYS is employed (ANSYS Inc.). In particular, ICEM CFD software is adopted as the preprocessing tool to build a mesh for the robot, which forms the finite flow domain, and Fluent is applied to simulate the flow and pressure distribution around the robot when it is in motion.

In the following, we take the dolphin body for example to introduce the whole CFD simulation process. For the flippers and flattened fluke, the CFD simulations adopt similar settings. In the preprocessing meshing work, an unstructured tetrahedron mesh is formed to describe the flow domain for great adaptability and high quality, as shown in Figure 2.16. The whole computation domain of the dolphin body is surrounded by the following boundaries:

1. *Inlet boundary*: Two times body lengths from the nose and set as velocity-inlet with $v = 0.1$ m/s.

2. *Outlet boundary*: Three times body lengths from the fluke and set as outflow.

3. *Top and bottom boundaries*: Set as velocity-inlet as $v = 0.1$ m/s in order to avoid reflected effects.

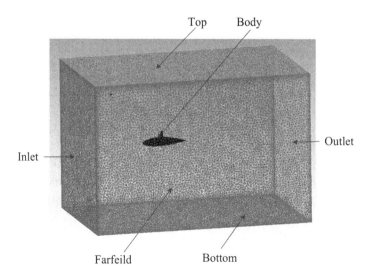

FIGURE 2.16   The boundary conditions for the gliding robotic dolphin.

4. Far-field boundary: Set as no-flip walls.

5. Surface boundary: Set as no-slip moving walls.

Meanwhile, for better simulation results, seven prismatic layers are stacked onto the surface mesh, as shown in Figures 2.17a and 2.17b. The CFD simulations about the flippers and flattened fluke adopt similar boundary conditions and several prismatic layers, as shown in Figures 2.17c and 2.17d. In addition, the fluid is supposed as an incompressible and steady one, and $k\text{-}w$ SST (shear–stress–transport) turbulence model with low-Re corrections is adopted in Fluent simulation.

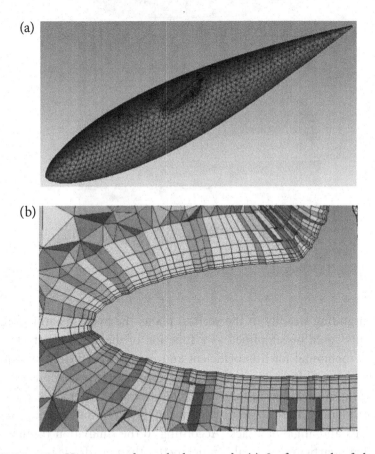

FIGURE 2.17 Unstructured tetrahedron mesh. (a) Surface mesh of the body. (b) Cut plan of volume mesh around the cylindrical body. (c) Surface mesh of the flipper. (d) Surface mesh of the flattened fluke and cut plan of volume mesh around the flattened fluke.

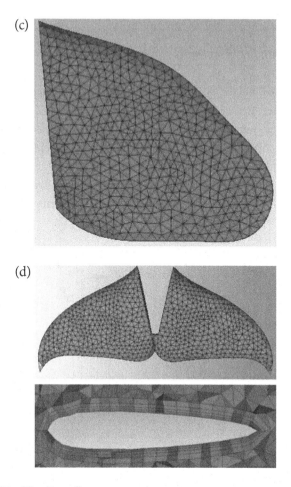

FIGURE 2.17 (*Continued*)

When gliding steadily in the vertical plane, the hydrodynamic forces coefficients could be simplified as a function of the angle of attack, for example, monomial for lift coefficient and moment coefficient and quadratic polynomial for a drag coefficient, according to the previous results [180]–[181]. For the gliding robotic dolphin, the asymmetric body shape leads to a slightly more complex relationship between the hydrodynamic coefficients and angle of attack. According to the simulation results, we found that quartic polynomial is much better than quadratic polynomial or monomial to fit the curves of the hydrodynamic coefficients of the dolphin body, as shown in Figure 2.18a–c. For the flippers and flattened flukes, the quadratic polynomial and monomial could obtain an expected

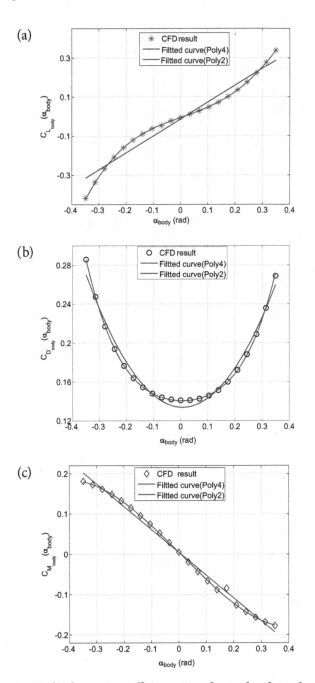

FIGURE 2.18 Hydrodynamic coefficients over the angle of attack. (a) Lift coefficient of the dolphin body. (b) Drag coefficient of the dolphin body. (c) Pitching moment coefficient of the dolphin body. (d) Lift coefficient of the flipper. (e) Drag coefficient of the flipper. (f) Pitching moment coefficient of the flipper.

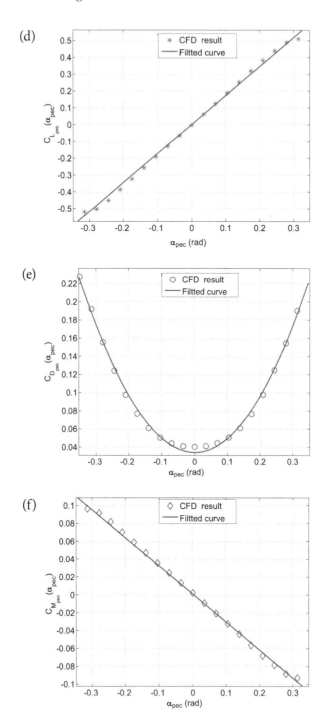

FIGURE 2.18 (*Continued*)

fitting result, as shown in Figure 2.18d–f. According to the earlier CFD simulation results, we can obtain the hydrodynamic coefficients for the dolphin body, flippers, and flattened fluke as follows:

$$\begin{cases} C_{D_{body}}(\alpha) = 6.489\alpha^4 + 0.0502\alpha^3 + 1.212\alpha^2 - 0.02245\alpha + 0.2588 \\ C_{L_{body}}(\alpha) = 0.5397\alpha^4 + 8.302\alpha^3 - 0.1858\alpha^2 + 0.7755\alpha - 0.01042 \\ C_{M_{body}}(\alpha) = 0.3971\alpha^4 + 0.4672\alpha^3 - 0.02648\alpha^2 - 0.214\alpha + 0.0011 \end{cases} \quad (2.2)$$

$$\begin{cases} C_{D_{pec}}(\alpha) = 1.481\alpha^2 - 0.000357\alpha + 0.6758 \\ C_{L_{pec}}(\alpha) = 1.667\alpha - 0.003487 \\ C_{M_{pec}}(\alpha) = -0.09995\alpha + 0.0009239 \end{cases} \quad (2.3)$$

$$\begin{cases} C_{D_{fluke}}(\alpha) = 1.344\alpha^2 - 0.002419\alpha + 0.09103 \\ C_{L_{fluke}}(\alpha) = 1.601\alpha \\ C_{M_{fluke}}(\alpha) = 0.3446\alpha - 0.000524 \end{cases} \quad , \quad (2.4)$$

where $C_{D_i}, C_{L_i}, C_{M_i}$ ($i = body, pec, fluke$) separately represent the hydrodynamic drag, lift, and moment coefficients by the cross-sectional area for dolphin body, flippers, and flattened fluke.

The analyses of the dynamic and static pressure distribution around the gliding robotic dolphin are also executed. Figure 2.19 separately displays

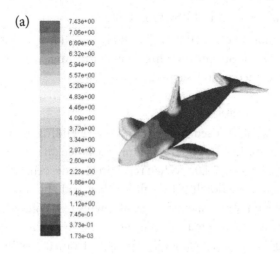

FIGURE 2.19 Pressure contour. (a) Dynamic pressure. (b) Static pressure.

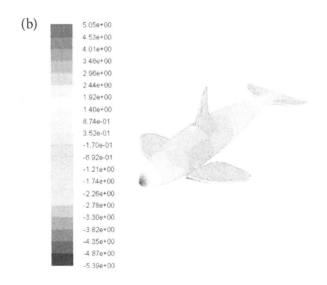

(b)

5.05e+00	
4.53e+00	
4.01e+00	
3.48e+00	
2.96e+00	
2.44e+00	
1.92e+00	
1.40e+00	
8.74e-01	
3.52e-01	
-1.70e-01	
-6.92e-01	
-1.21e+00	
-1.74e+00	
-2.26e+00	
-2.78e+00	
-3.30e+00	
-3.82e+00	
-4.35e+00	
-4.87e+00	
-5.39e+00	

FIGURE 2.19 (*Continued*)

pressure distribution around the robot. We can see that the highest pressure is at the tip of the robot's nose and the windward side of every fin or fluke, due to the interreaction between the fluid and the gliding robotic dolphin. These pressures on the rest of the robot surface are both lower due to the smooth flow. Generally, these pressures are all so small to be sustainable and not bring some destruction effect for the gliding robotic dolphin.

2.4 EXPERIMENTS AND DISCUSSION

To validate the motion capability of the developed gliding robotic dolphin for both gliding motion and dolphin-like swimming, extensive experiments were performed in this section.

2.4.1 A Miniature Prototype

The first experiment focused on the downward gliding motion. At the beginning, the dolphin-like glider got a force balance on the surface of the water, as shown in Figure 2.20a. When receiving the descending command from the computer, the dolphin-like glider absorbed about 4.6 g of water to change its buoyancy and started to glide downward (Figure 2.20b–i). To explore how the angle affected the gliding speed, the turn angle between flippers and the body was manually changed from 0° to 15° every 5°. Figure 2.21 gives the relationship between downward gliding velocities

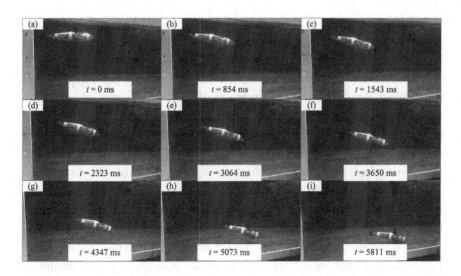

FIGURE 2.20   Snapshot sequence of downward gliding motion. (a) $t = 0$ ms,
(b) $t = 854$ ms, (c) $t = 1543$ ms, (d) $t = 2323$ ms, (e) $t = 3064$ ms, (f) $t = 3650$ ms,
(g) $t = 4347$ ms, (h) $t = 5073$ ms, (i) $t = 5811$ ms.

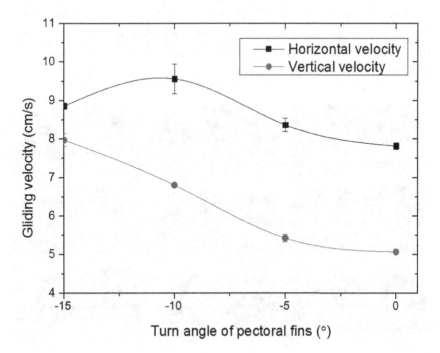

FIGURE 2.21   The relationship between downward gliding velocities and the
turn angle of flippers.

and the turn angle of flippers. From Figure 2.21, we can see that the dolphin-like glider gained the highest horizontal speed of 9.56 cm/s when $\beta = 10°$ and the highest vertical speed of 7.97 cm/s when $\beta = 15°$.

The second experiment was carried out to testify to the upward gliding motion. Similar to the downward motion, the upward gliding motion could be successfully realized through draining away 4.6 g of water, as shown in Figure 2.22. According to the experimental results, we can see that the dolphin-like glider could obtain the highest horizontal speed of 6.27 cm/s when $\beta = 5°$ and the highest vertical speed of 4.96 cm/s when 15°. We can see that the dolphin-like glider had different vertical speeds in upward and downward gliding motion, although β had the same value (–15° and 15°). This phenomenon is mainly due to the asymmetric body shape of the dolphin-like glider, which leads to different hydrodynamic performances such as different pitch angles and gliding path angles in upward gliding motion and downward gliding motion. Therefore, the dolphin-like glider obtained different vertical speeds in upward and downward gliding motions, although the turn angles of flippers had the same value (–15° and 15°; Figure 2.23).

Traditional underwater gliders usually use the internal moveable masses to regulate the gliding attitude. Because of the back-and-forth movements, the internal masses often occupy large spaces, which lead to low space utilization rates. Comparatively, the dolphin-like glider provided in this

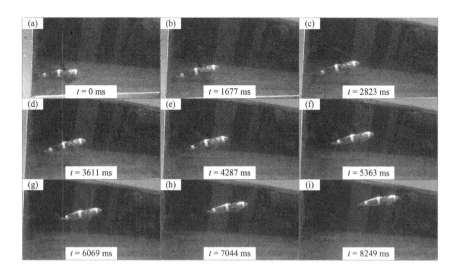

FIGURE 2.22   Snapshot sequence of upward gliding motion. (a) $t = 0$ ms, (b) $t = 1677$ ms, (c) $t = 2823$ ms, (d) $t = 3611$ ms, (e) $t = 4287$ ms, (f) $t = 5363$ ms, (g) $t = 6069$ ms, (h) $t = 7044$ ms, (i) $t = 8249$ ms.

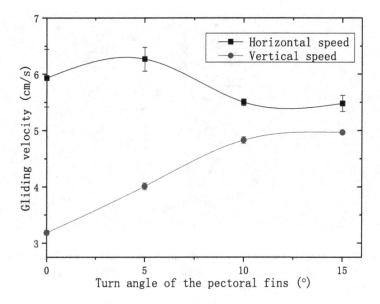

FIGURE 2.23 The relationship between upward gliding velocities and the turn angle of flippers.

section could obtain enough pitching moments from both the buoyancy-driven system and controllable fins, including flippers and flattened flukes. Moreover, the flukes often provide a considerable pitching moment due to the relative larger moment arm. In this situation, the buoyancy-driven system only needs a little volume for water, about ±0.6% of the whole displacement. The volume of the buoyancy-driven system and turn angle of the controllable fins could also be used as controlled input variables for an expected accurate attitude. Moreover, flexible flippers and flattened flukes could bring quick response capability into the attitude adjustment.

2.4.2 A 1-m-Scale Prototype

The first experiment focused on the gliding motion. When receiving the descending command, the gliding robotic dolphin began to absorb the water to change its buoyancy and the flippers synchronously turn a positive angle to provide some additional pitch torques. After several seconds, the gliding robotic dolphin adjusted to gliding upward by pumping the water and turning a negative angle for the flippers, as shown in Figure 2.24f. In order to obtain the gliding performance, several experiments were executed under the different turn angles of flippers, for example, 15° and 30°, see Table 2.3. Finally, the gliding robotic dolphin

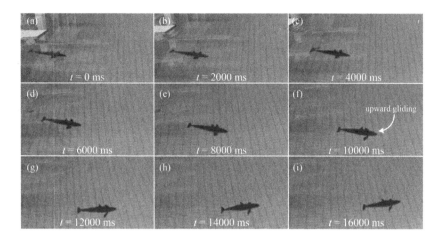

FIGURE 2.24   Snapshot sequence of the gliding motion. (a) $t = 0$ ms, (b) $t = 2000$ ms, (c) $t = 4000$ ms, (d) $t = 6000$ ms, (e) $t = 8000$ ms, (f) $t = 10000$ ms, (g) $t = 12000$ ms, (h) $t = 14000$ ms, (i) $t = 16000$ ms.

TABLE 2.3   The Gliding Velocity of the Gliding Robotic Dolphin

φpec (°)	15				30			
No.	#1	#2	#3	AVG	#1	#2	#3	AVG
Vx (cm/s)	15.98	15.74	14.77	15.50	13.68	14.71	13.14	13.84
Vz (cm/s)	13.79	14.05	13.36	13.73	16.48	17.24	16.22	16.65
Gliding angle (°)	40.83	41.78	42.17	41.49	50.33	49.55	51.02	50.30

successfully realized the gliding motion with about 0.155 m/s horizontal gliding speed when $\varphi_{pec} = 15°$.

The second experiment focuses on the dolphin-like swimming, including MPF locomotion and BCF locomotion, as shown in Figures 2.25 and 2.26. In order to produce efficient and effective dolphin-like swimming, a central pattern generator (CPG) model based on Hopf oscillators was adopted. The details in the control method can be referred to our previous work [182], [183]. In BCF locomotion, the body CPGs are activated to generate the rhythmic signals, while the flippers' CPGs are inactivated to keep the flippers horizontal. Similar to our previous results, the robotic dolphin could successfully realize forward swimming and flexible turn, and the propulsive speed increases directly with the flapping frequency. Moreover, the robotic dolphin could realize a spiraling motion in MPF locomotion. Figure 2.25 depicts a slow spiraling motion of the robotic dolphin. In this experiment, the gliding robotic dolphin obtained a left turn torque if the

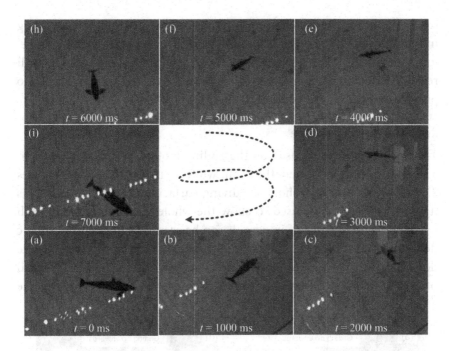

FIGURE 2.25 Snapshot sequence of the spiraling motion by the flippers. (a) $t = 0$ ms, (b) $t = 1000$ ms, (c) $t = 2000$ ms, (d) $t = 3000$ ms, (e) $t = 4000$ ms, (f) $t = 5000$ ms, (g), (h) $t = 6000$ ms, (i) $t = 7000$ ms.

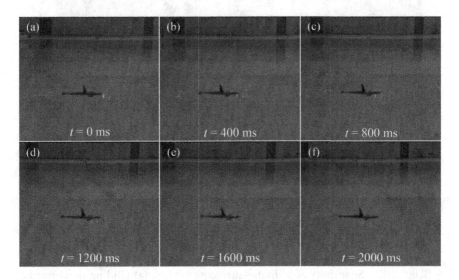

FIGURE 2.26 Snapshot sequence of the dolphin-like motion. (a) $t = 0$ ms, (b) $t = 400$ ms, (c) $t = 800$ ms, (d) $t = 1200$ ms, (e) $t = 1600$ ms, (f) $t = 2000$ ms.

left flipper has a 5° lower turn angle than the right one. Then, a descending motion can also be realized if both flippers have a positive turning angle, for example, 30° in this experiment. Moreover, the robotic dolphin will reach a better spiraling motion if the buoyancy-driven system joins in to provide descending thrusts and moments.

2.4.3 A 1.5-m Prototype

The first experiment focused on the gliding motion, which was tested in a diving training pool (length: 16 m, width: 6 m, depth: 7.5 m). Figures 2.27a–c present the snapshots of a diving-surfacing period. Note that this figure was simply processed to show the whole gliding course through overlapping a sequence of images. It can be seen that the gliding robotic dolphin successfully realized a sawtooth-like motion. In fact, the robotic dolphin hardly reaches the equilibrium state, with all the oil being sucked or ejected in a diving-surfacing period, because of the limited size of the

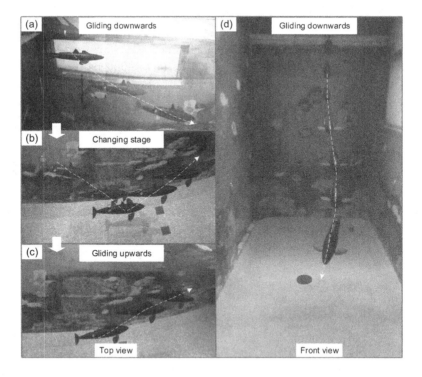

FIGURE 2.27 Snapshots of a diving-surfacing period. (a) Gliding downwards, (b) Changing stage, (c) Gliding upwards (Top view), (d) Gliding downwards (Front view).

pool and poor speed of the buoyancy-driven system. In the experiment, the pump rotated at 50 r/s from the beginning to 21 s, and the movable mass shifted 3.9 cm ahead at 3 mm/s. Meanwhile, the robotic dolphin started to glide downward slowly under the action of the net buoyancy. When the diving depth exceeded 2 m, the robotic dolphin should switch to glide upward according to the precommand from embedded programs. At this point, the pump worked again at 35 r/s. Thanks to these onboard sensors including the inertial measurement unit and the pressure sensor, we could collect real-time attitude and depth data during the whole process and compare the simulated and experimental data, which correspond to a diving-surfacing period. According to the collected depth data, the actual switching depth was 4.4 m. Apparently, the robotic dolphin spent very long time switching the gliding state from diving to surfacing because of a slow change of the net buoyancy. In order to seek the equilibrium states, we tried to release the robot after all the oil was sucked. Several experiments were executed to obtain the gliding performance. Table 2.4 illustrates the statistical summary of our test results on gliding motion. Note that μ and σ denote the mean and variance of every statistical index, respectively. As can be observed, the average horizontal speed during a steady gliding state reaches about 0.356 m/s. On the other hand, the average gliding angle is approximately $-36.2°$, which is close to that of traditional underwater gliders. This illustrates that the robotic dolphin modeling after a killer whale can perform an expected gliding motion as traditional gliders, although it has a smaller wingspan. Based on the average gliding angle and pitch angle, we can further obtain the steady angle of attack of the robot, about $-12°$. Under this angle of attack, the gliding robotic dolphin has a relative higher lift–drag ratio, which can contribute to a great gliding efficiency.

By virtue of the flippers, the robot can perform a 3-D spiraling motion. To yield a spiraling case, the left flipper maintains a certain deflection

TABLE 2.4 Summary of the Test Results of the Gliding Motion

No.	x (m)	Vx (m/s)	Vz (m/s)	Gliding angle (°)	Pitch angle (°)
#1	11.3	0.404	0.269	−33.7	−20.2
#2	9.5	0.343	0.271	−38.3	−28.5
#3	10.1	0.346	0.249	−36.6	−25.5
#4	10.2	0.440	0.250	−36.3	−22.3
μ	10.3	0.356	0.260	−36.2	−24.1
σ	0.7	0.032	0.012	1.9	3.6

angle, while the other actuators execute the same actions with the 2-D gliding case earlier for diving. Therefore, we also conducted experiments of the spiraling motion, as shown in Figure 2.28. To make sure the spiraling motion arrive as soon as possible, we also released the robotic dolphin after all the oil was sucked and maintained the left flipper a deflection angle of 60°. It should be noted that some shocks on the robot in the beginning of the release stage caused the fluctuation of roll and pitch angles.

The second experiment concerned the dolphin-like swimming, including BCF locomotion and MPF locomotion. In BCF locomotion, the gliding robotic dolphin drove the posterior body with flatten flukes to pitch up and down, but the flippers kept still, indicating that the waist and cauda CPG units are activated, and the flipper CPG units are inactivated to keep the flippers horizontal. As for the waist and cauda CPG units, the amplitudes were set to 25° and 35°, respectively, and the phase lag was set to 60°. With an increase in flapping frequency, the gliding robotic dolphin could swim much faster. Figure 2.29 illustrates the dolphin-like swimming in the BCF

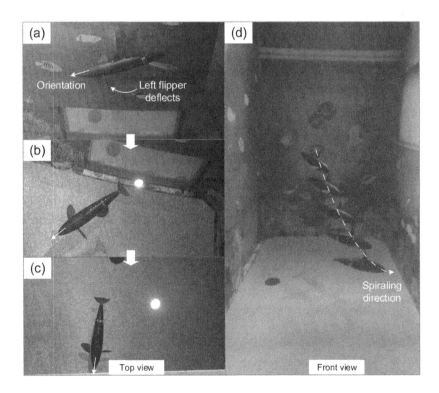

FIGURE 2.28   Snapshots of a spiraling motion. (a)–(c) Top view and (d) front view.

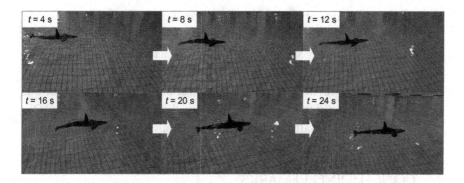

FIGURE 2.29   Snapshots of the sequence of the dolphin-like motion.

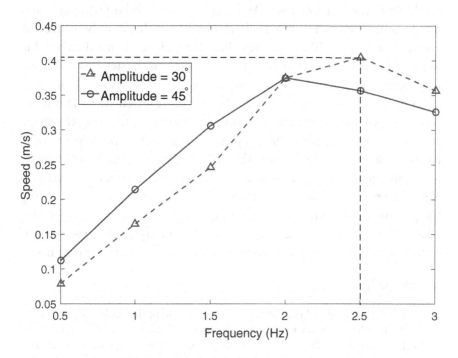

FIGURE 2.30   Plot of the measured propulsive speed in the MPF locomotion.

locomotion. The robotic dolphin obtained steady dolphin-like swimming, and the propulsive speed reached 0.506 m/s under an undulation frequency of 0.8 Hz. In the MPF locomotion, two groups of amplitudes including 30° and 45° were set for the flipper CPG units. As shown in Figure 2.30, the propulsive speed also increased with the flapping frequency. Clearly, a larger flapping frequency will lead to a decreased propulsive speed since the flippers cannot follow the expected amplitudes. Finally, the MPF locomotion attained a steady speed of 0.405 m/s with a frequency of 2.5 Hz and an amplitude of 30°. Furthermore, another experiment was also carried out to examine the turning capability of the robot in the MPF locomotion. On the basis of the differential motions of two flippers with separate DOF, the robotic dolphin finally realized the turn maneuvers with a speed of 7.8°/s and a turning radius of 1.7 BLs.

2.5 CONCLUDING REMARKS

In this chapter, we aim at an innovative concept for the gliding underwater robots, namely, gliding robotic dolphins, to implement both high maneuverability and long endurance. As a combination of robotic dolphins and underwater gliders, the gliding robotic dolphin could not only realize fast and flexible dolphin-like swimming but also glide for a long distance due to the buoyancy-driven system. First, we presented the detailed mechanical design of four newly gliding robotic dolphins, which differ from each other in size and mechanical design. Furthermore, extensive CFD simulations were conducted, which not only can assist in optimizing the shape design to improve gliding efficiency but also can obtain the key hydrodynamic coefficients to support the theoretical research. Finally, on the basis of the designed prototypes, extensive aquatic experiments and analyses were carried out to show the multimodal motion performances, including the gliding and dolphin-like motion modes. To sum up, our works offer a valuable idea for future underwater robots, and the gliding robotic dolphin has the significant potential to work in complex ocean environments.

REFERENCES

[168] J. J. Rohr and F. E. Fish, "Strouhal numbers and optimization of swimming by odontocete cetaceans," *J. Exp. Biol.*, vol. 207, no. 10, pp. 1633–1642, 2004.

[169] I. Nesteruk, G. Passoni, and A. Redaelli, "Shape of aquatic animals and their swimming efficiency," *J. Mar. Biol.*, vol. 2014, 2014, Art. no. 470715.

[170] F. E. Fish, P. Legac, T. M. Williams, and T. Wei, "Measurement of hydrodynamic force generation by swimming dolphins using bubble DPIV," *J. Exp. Biol.*, vol. 217, no. 2, pp. 252–260, 2014.

[171] F. E. Fish and J. J. Rohr, "Review of dolphin hydrodynamics and swimming performance," U.S. Navy, San Diego, CA, USA, Tech. Rep. 1801, Aug. 1999.

[172] J. Liu, Z. Wu, J. Yu, and M. Tan, "Sliding mode fuzzy control-based path-following control for a dolphin robot," *Sci. China Inf. Sci.*, vol. 61, no. 2, 024201, 2018.

[173] Y. Park, T. Huh, D. Park, and K. Cho, "Design of a variable-stiffness flapping mechanism for maximizing the thrust of a bio-inspired underwater robot," *Bioinsp. Biomim.*, vol. 9, no. 3, Art. no. 036002, 2014.

[174] M. Nakashima, T. Tsubaki, and K. Ono, "Three-dimensional movement in water of the dolphin robot-control between two positions by roll and pitch combination," *J. Robot. Mechatron.*, vol. 18, no. 3, pp. 347–355, 2006.

[175] G. Ren, Y. Dai, Z. Cao, and F. Shen, "Research on the implementation of average speed for a bionic robotic dolphin," *Robot. Auton. Syst.*, vol. 74, Part A, pp. 184–194, 2015.

[176] J. Yu, Z. Su, Z. Wu, and M. Tan, "Development of a fast-swimming dolphin robot capable of leaping," *IEEE/ASME Trans. Mechatronics*, vol. 21, no. 5, pp. 2307–2316. 2016.

[177] H. Stommel, "The slocum mission," *Oceanography*, vol. 2, no. 1, pp. 22–25, 1989.

[178] N. E. Leonard, D. A. Paley, R. E. Davis, D. M. Fratantoni, F. Lekien, and F. Zhang, "Coordinated control of an underwater glider fleet in an adaptive ocean sampling field experiment in Monterey Bay," *J. Field Robot.*, vol. 27, no. 6, pp. 718–740, 2010.

[179] AID. http://www.airfoildb.com/foils/376

[180] S. Zhang, J. Yu, A. Zhang, and F. Zhang, "Spiraling motion of underwater gliders: Modeling, analysis, and experimental results," *Ocean Eng.*, vol. 60, no. 1, pp. 1–13, 2013.

[181] B. Zhao, R. Skjetne, M. Blanke, and F. Dukan, "Particle filter for fault diagnosis and robust navigation of underwater robot," *IEEE Trans. Control Syst. Technol.*, vol. 22, no. 6, pp. 2399–2407, 2014.

[182] Z. Wu, J. Yu, Z. Su, and M. Tan, "An improved multimodal robotic fish modelled after Esox lucius," in *Proc. IEEE Int. Conf. Robot. Biomim.*, Shenzhen, China, Dec. 2013, pp. 516–521.

[183] Z. Wu, J. Yu, M. Tan, and J. Zhang, "Kinematic comparison of forward and backward swimming and maneuvering in a self-propelled sub-carangiform robotic fish," *J. Bionic Eng.*, vol. 11, pp. 199–212, 2014.

3-D Motion Modeling of the Gliding Underwater Robot

3.1 INTRODUCTION

Furthermore, regarding the gliding underwater robot, this chapter establishes the dynamic model and offers some comparisons of the spiraling motion based on the two different swimming modes accompanied by corresponding comparative analyses. Finally, we verify the complete dynamic model and analyze the yaw angle and diving depth by comparing the aquatic experiments with simulations.

Further progress is made in this chapter, whose objective is to establish a complete dynamic model followed by the motion analysis. The contributions made in this chapter are summarized as follows:

- We establish a complete dynamic model with the full consideration of gliding motion and dolphin-like motions. With regard to gliding motion, the equations are derived here by computing momenta from the total vehicle-fluid system energy. As for the dolphin-like motion, we apply the multilink dynamics.

- For gliding motion, we offer the motion performance analysis a gliding underwater robot based on the established dynamic model.

DOI: 10.1201/9781003347439-3

- Extensive simulations and aquatic experiments are conducted. The obtained results validate the effectiveness and accuracy of the dynamic model.

The remainder of this chapter is organized as follows: Section 3.2 provides the motion model of the gliding underwater robot, including the kinematic model, and dynamic models of the gliding motion and dolphin-like motion. Section 3.3 presents the motion performance analysis of the gliding underwater robot in a gliding motion. Section 3.4 shows the experimental results followed by a discussion. Section 3.5 concludes this chapter with an outline of future work.

3.2 MOTION MODELING OF THE GLIDING UNDERWATER ROBOT

To clearly describe a complete dynamic model of the gliding underwater robot, coordinate systems including an inertial frame, a body-fixed frame, a waist frame, a tail frame, and two flippers' frames are defined first. All the coordinate frames are illustrated in Figure 3.1, and these frames follow the right-hand rule. Moreover, we denote the inertia frame $C_g = o_g x_g y_g z_g$, the z-axis of which is along the direction of gravity, while the x and y are perpendicular to z. Next, we define a body-fixed frame $C_b = o_b x_b y_b z_b$, the origin of which locates at the center of buoyancy (CB). In particular, $C_w = o_w x_w y_w z_w$, $C_t = o_t x_t y_t z_t$, $C_l = o_l x_l y_l z_l$, and $C_r = o_r x_r y_r z_r$ represent the joint frames of waist, tail, left flipper, and right flipper, respectively. Furthermore, we define $j = [0,1,0]^T$, $k = [0,0,1]^T$, and $J = [0_{1\times3}, j]^T$. Finally, for two vectors $p, q \in R^3$, their cross-product is denoted as $p \times q = \hat{p} \cdot q$, where $\hat{p}$ represents the skew matrix of p.

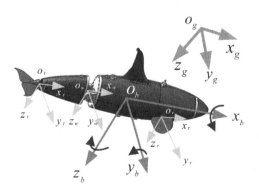

FIGURE 3.1 Coordinate systems including inertial, body, and fin frames.

3.3 KINEMATIC ANALYSIS

The kinematic equations of the gliding underwater robot include body kinematics and fin surface kinematics, which represent the speed information in the body coordinate system and each fin surface coordinate system, respectively. Denote $U_b = (U_{bx}, U_{by}, U_{bz})^T$ and $\Omega_b = (\Omega_{bx}, \Omega_{by}, \Omega_{bz})^T$ the line velocity and angular velocity of the gliding underwater robot with respect to (w.r.t.) the body frame, respectively. Therefore, $V_b = (U_b{}^T, \Omega_b{}^T)^T$ denotes the velocity vector. The kinematics of the robot are formalized by

$$
\begin{aligned}
{}^g \dot{P}_b &= {}^g U_b = {}^g R_b U_b \\
{}^g \dot{R}_b &= {}^g R_b \hat{\Omega}_b
\end{aligned}, \tag{3.1}
$$

where ${}^g R_b$ and ${}^g P_b$ denote the rotation matrix and position vector of C_b w.r.t. C_g, and ${}^g R_b$ is determined by roll angle ψ, pitch angle θ, and yaw angle φ. Denote c, s, t be cos, sin, and tan, ${}^g P_b$ can be calculated as follows:

$$
{}^g R_b = \begin{pmatrix} c\psi c\theta & -s\psi c\phi + c\psi s\theta s\phi & s\psi s\phi + c\psi c\phi s\theta \\ s\psi c\theta & c\psi c\phi + s\psi s\theta s\phi & -c\psi s\phi + s\theta s\psi c\phi \\ -s\theta & c\theta s\phi & c\theta c\phi \end{pmatrix} \tag{3.2}
$$

By defining the velocity vectors w.r.t. the global framework $V_g = (U_g{}^T, \Omega_g{}^T)^T$ and w.r.t. the body framework $V_b = (U_b{}^T, \Omega_b{}^T)^T$, the kinematic model can be derived as follows:

$$
\Omega_g = J\Omega_b, \tag{3.3}
$$

where

$$
J = \begin{pmatrix} 1 & s\phi t\theta & c\phi t\theta \\ 0 & c\phi & -s\phi \\ 0 & s\phi / c\theta & c\phi / c\theta \end{pmatrix}.
$$

Afterward, since the surfaces driven by flipper joints, the waist joint, and the tail joint are movable, we should consider their offset angles for their kinematics:

$$
\begin{aligned}
V_i &= {}^i H_b V_b + \delta_i \qquad (i = w, l, r) \\
V_t &= {}^t H_w V_w + \delta_t
\end{aligned}, \tag{3.4}
$$

where

$$
{}^{i}H_{b} = {}^{b}H_{i}{}^{T} = \begin{pmatrix} {}^{i}R_{b} & -{}^{i}R_{b}{}^{b}\hat{P}_{i} \\ 0_{3\times 3} & {}^{i}R_{b} \end{pmatrix} \quad (i=w,l,r)
$$

$$
{}^{t}H_{w} = {}^{w}H_{t}{}^{T} = \begin{pmatrix} {}^{t}R_{w} & -{}^{t}R_{w}{}^{w}\hat{P}_{t} \\ 0_{3\times 3} & {}^{t}R_{w} \end{pmatrix}
$$

$$
\delta_{i} = \dot{\theta}_{i}J_{i} \quad (i=w,t,l,r).
$$

The indexes b, w, t, l, r correspond to the body, waist, tail, left flipper, and right flipper, respectively. δ_{i} indicates the speed change caused by the fin surface's movements. ${}^{b}P_{i}(i=w,l,r)$ represents a corresponding position vector, and ${}^{i}R_{b}(i=w,l,r)$ indicates the rotation matrix from coordinate frame C_{b} to C_{i}, which is related to offset angle θ_{i} of movable surfaces w.r.t. frame C_{b}.

$$
{}^{i}R_{b} = {}^{b}R_{i}{}^{T} = \begin{pmatrix} \cos\theta_{i} & 0 & -\sin\theta_{i} \\ 0 & 1 & 0 \\ \sin\theta_{i} & 0 & \cos\theta_{i} \end{pmatrix} \tag{3.5}
$$

Regarding the rotation matrix and position vectors from frame C_{w} to C_{t}, there exists the same form and meaning.

3.3.1 Net Buoyancy Analysis

One of main external forces is the net buoyancy, which represents the difference between gravity and buoyancy. When the position of the piston is at the middle point of water injector, the net buoyancy is zero. Therefore, the net buoyancy of the robot w.r.t. frame C_{b} is given by

$$
G_{n} = \rho S(\frac{h_{o}}{2} - h)g({}^{g}R_{b}{}^{T}k), \tag{3.6}
$$

where ρ indicates the density of the water. g denotes the gravitational acceleration. S and h_{o} indicate the bottom area and the total height of the water injector, respectively. h denotes the real position of the piston. Hence, the moment of net buoyancy takes the forms as follows:

$$
\tau_{n} = (m_{b}\hat{P}_{b} + m_{s}\hat{P}_{s} + m_{m}\hat{P}_{m})g({}^{g}R_{b}{}^{T}k) + G_{n}\hat{P}_{in}, \tag{3.7}
$$

where m_b is the body's mass excluding the movable and water injector mass. m_m and m_s denote the movable and water injector mass, respectively. P_b, P_m, and P_s are the position vectors of corresponding center of gravity (CG) w.r.t. frame C_b. P_{in} indicates the movable vector of CG caused by the movement of the piston. Hence, we define $G_b = (G_n, \tau_n)^T$.

3.3.2 Hydrodynamic Analysis

Hydrodynamics is another major external force, which is analyzed with the quasi-steady model in this chapter. Hydrodynamic is closely related to the relative attitude that can be parameterized by the angle of attack α_i and the sideslip angle $\beta_i (i = b, w, t, l, r)$. For convenience, we introduce a velocity coordinate frame $C_v = o_v x_v y_v z_v$ to characterize the relative attitude, as shown in Figure 3.2. Therefore, the hydrodynamic forces of body and movable surfaces and their moments can be calculated by

$$
{}^v F_i = \begin{pmatrix} -{}^v D_i \\ {}^v SF_i \\ -{}^v L_i \end{pmatrix} = \frac{1}{2} \rho S_i U_i^2 \begin{pmatrix} -C_{i,d}(\alpha_i) \\ C_{i,sf}(\beta_i) \\ -C_{i,l}(\beta_i) \end{pmatrix} \tag{3.8}
$$

$$
{}^v \tau_b = \begin{pmatrix} {}^v \tau_{ix} \\ {}^v \tau_{iy} \\ {}^v \tau_{iz} \end{pmatrix} = \frac{1}{2} \rho S_i U_i^2 \begin{pmatrix} C_{i,\tau x}(\beta_i) \\ C_{i,\tau y}(\alpha_i) \\ C_{i,\tau z}(\beta_i) \end{pmatrix} + K_i \Omega_i, \tag{3.9}
$$

where

$$
\alpha_i = \arctan\left(\frac{U_{iz}}{U_{ix}}\right) \qquad \beta_i = \arcsin\left(\frac{U_{iy}}{\|U_i\|}\right)
$$

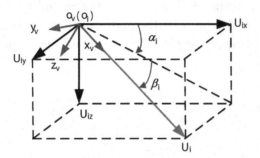

FIGURE 3.2 Velocity coordinate frame.

and S_i presents the reference area of body and movable surfaces. C indicates the corresponding hydrodynamic coefficients related to the angle of attack and the sideslip angle. K_i denotes the matrix of the rotating damping coefficients.

Next, since the forces of body and movable surfaces are expressed in the velocity coordinate frame, it is necessary to transform them to their own coordinate frames with the rotation matrix ^iR_v. Furthermore, we unify all the physical quantities into the body coordinate frame

$$\begin{pmatrix} ^bF_i \\ ^b\tau_i \end{pmatrix} = \begin{pmatrix} ^bR_i & 0_{3\times3} \\ ^b\hat{P}_i^{\,b}R_i & ^bR_i \end{pmatrix}\begin{pmatrix} ^iR_v{}^vF_i \\ ^iR_v{}^v\tau_i \end{pmatrix}, \tag{3.10}$$

where $i = b,w,t,l,r$, $^bR_b = I_{3\times3}$, and $^b\hat{P}_b = 0_{3\times3}$.

3.3.3 Dynamic Model

For the convenience of analysis, this section divides the dynamic analysis of the gliding underwater robot into a gliding module and a dolphin-like module.

With regard to gliding motion, the equations are derived here by computing momenta from the total vehicle-fluid system energy. Let p and π denote the total translational and total angular momentum of the system, respectively, which are all expressed w.r.t. inertial frame. Let P and Π represent the momentum expressed w.r.t. body frame. Therefore, the conversion relationship between them is as follows:

$$\begin{aligned} p &= RP \\ \pi &= R\Pi + l \times p, \end{aligned} \tag{3.11}$$

Where l is the vector from the origin of the inertial frame to the origin of the body frame. Furthermore, denote T as the system's total kinetic energy, and $T_i (i = b, f, m, s)$ represents the body, additional, movable mass, and injector kinetic energy, respectively. The total kinetic energy T is computed as follows:

$$\begin{aligned} T = T_b + T_f + T_m + T_s &= \frac{1}{2}\begin{pmatrix} U_b^T & \Omega_b^T \end{pmatrix}\tilde{M}\begin{pmatrix} U_b \\ \Omega_b \end{pmatrix} \\ &+ \frac{1}{2}\sum_{i=m,s}\left(U_b + \dot{r}_i + \Omega_b \times r_i\right)^T m_i\left(U_b + \dot{r}_i + \Omega_b \times r_i\right) \end{aligned} \tag{3.12}$$

where

$$\tilde{M} = \begin{pmatrix} m_b I + M_f & (m_b \hat{r}_b + D_f)^T \\ m_b \hat{r}_b + D_f & J_b + J_f + m_b \hat{r}_b^T \hat{r}_b \end{pmatrix}.$$

In order to distinguish between position vectors and momentum variables, we attach the $r_i (i = b, m, s)$ to the position of the i part in the body-fixed frame. The $m_i (i = b, f, m, s)$ represents the mass of the i part. J_b is the rotational inertial for the uniformly distributed m_b. Furthermore, it is known that the dots of translational and total angular momentum are external force and external moment, respectively. Hence, the dots take the forms as follows:

$$\dot{P} = P \times \Omega_b + R^T \sum_{i=1}^{I} f_{ext_i}$$

$$\dot{\Pi} = \Pi \times \Omega_b + P \times U_b + R^T \sum_{j=1}^{J} \tau_{ext_j}$$

(3.13)

The translational and total angular momentum can be obtained as

$$P = \frac{\partial T}{\partial U_b} = \left(m_b I + M_f\right)U_b + (m_b \hat{r}_b + D_f)^T \Omega_b$$
$$+ \sum_{i=m,s} m_i \left(U_b + \dot{r}_i + \Omega_b \times r_i\right)$$

(3.14)

$$\Pi = \frac{\partial T}{\partial \Omega_b} = \left(J_b + J_f + m_b \hat{r}_b^T \hat{r}_b\right)\Omega_b + \left(m_b \hat{r}_b + D_f\right)U_b$$
$$+ \sum_{i=m,s} m_i \hat{r}_i \left(U_b + \dot{r}_i + \Omega_b \times r_i\right).$$

(3.15)

Furthermore, the derivatives of the previously mentioned momentum and angular momentum can be obtained:

$$\dot{P} = \left(m_b I + M_f\right)\dot{U}_b + (m_b \hat{r}_b + D_f)^T \dot{\Omega}_b$$
$$+ \sum_{i=m,s} m_i \left(\dot{U}_b + \ddot{r}_i + \dot{\Omega}_b \times r_i + \Omega_b \times \dot{r}_i\right)$$

(3.16)

$$\dot{\Pi} = \left(J_b + J_f + m_b \hat{r}_b^T \hat{r}_b\right)\dot{\Omega}_b + \left(m_b \hat{r}_b + D_f\right)\dot{U}_b$$
$$+ \sum_{i=m,s} m_i \hat{r}_i \left(U_b + \dot{r}_i + \Omega_b \times r_i\right) + \sum_{i=m,s} m_i \hat{r}_i \left(\dot{U}_b + \ddot{r}_i + \dot{\Omega}_b \times r_i + \Omega_b \times \dot{r}_i\right).$$

(3.17)

For the convenience of analysis and presentation, some parameters are defined:

$$\tilde{M}_b = \sum_{j=b,m,s} m_i I + M_f$$

$$D = \sum_{j=b,m,s} m_i \hat{r}_i + D_f \qquad \qquad (3.18)$$

$$J = J_b + J_f + \sum_{j=b,m,s} m_i \hat{r}_i^T \hat{r}_i$$

Furthermore, substitute Equations 3.14 and 3.15 and its dot into Equations 3.16 and 3.17, we can obtain two equations for gliding motion:

$$\tilde{M}_b \dot{U}_b + D^T \dot{\Omega}_b = \left(\tilde{M}_b U_b \right) \times \Omega_b + \left(D^T \Omega_b \right) \times \Omega_b$$

$$+ \sum_{i=m,s} m_i \left(2\dot{r}_i \times \Omega_b - \ddot{r}_i \right) + R^T \sum_{i=1}^{I} f_{ext_i} \qquad (3.19)$$

$$D \dot{U}_b + J \dot{\Omega}_b = \left(\tilde{M}_b U_b \right) \times U_b + \left(D^T \Omega_b \right) \times U_b + \left(D U_b \right) \times \Omega_b + \left(J \Omega_b \right)$$

$$\times \Omega_b + \sum_{i=m,s} m_i \left(\hat{r}_i \left(2\hat{r}_i \Omega_b - \ddot{r}_i \right) \right) + R^T \sum_{k=1}^{K} \tau_{ext_j}. \qquad (3.20)$$

Arrange the preceding formula as follows:

$$\tilde{M}_b \begin{pmatrix} \dot{U}_b \\ \dot{\Omega}_b \end{pmatrix} = -\Gamma_{cb} + \Gamma_m + \Gamma_s + \Gamma_{extb}. \qquad (3.21)$$

Since the buoyancy and center of gravity adjustment mechanisms remain stationary most of the time during the gliding motion, it can be assumed that $\ddot{r}_m = \ddot{r}_s = 0_{3\times1}$. The speed of motion of the two mechanisms can be defined as

$$\dot{r}_m = U_m$$
$$\dot{r}_s = U_s \qquad \qquad (3.22)$$

The preceding derivation has obtained the dynamic model of the gliding underwater robot, buoyancy adjustment and gravity center adjustment mechanism, and its total external force (moment) Γ_{extb} includes hydrodynamic force (moment), net buoyancy (moment) and joints force (moment), among others. The dynamic equation of each movable fin can be obtained by a similar analysis method. As mentioned earlier, in this section, the gliding underwater robot is regarded as a multilink system, and the Newton–Euler method is used for dynamic analysis to obtain the dynamic equations of the following modules:

$$
\begin{aligned}
M_b \dot{V}_b &= -\Gamma_{cb} + \Gamma_{hb} + \Gamma_{wb} + \Gamma_{lb} + \Gamma_{rb} + G_b + \Gamma_m + \Gamma_s \\
{}^b H_w M_w \dot{V}_w &= {}^b H_w (-\Gamma_{cw} + \Gamma_{hw} + \Gamma_{bw} + \Gamma_{tw}) \\
{}^b H_t M_t \dot{V}_t &= {}^b H_t (-\Gamma_{ct} + \Gamma_{ht} + \Gamma_{wt}), \\
{}^b H_l M_l \dot{V}_l &= {}^b H_l (-\Gamma_{cl} + \Gamma_{hl} + \Gamma_{bl}) \\
{}^b H_r M_r \dot{V}_r &= {}^b H_r (-\Gamma_{cr} + \Gamma_{hr} + \Gamma_{br})
\end{aligned}
\tag{3.23}
$$

where $M_i (i = b,w,t,l,r)$ represents the total inertia matrix. Γ_m and Γ_s denote the forces of movable mass and water injector on body. $F_{ci} = (f_{ci}, \tau_{ci})^T (i = b,w,t,l,r)$ denotes the Coriolis force and moment on part i. $F_{hi} = (f_{hi}, \tau_{hi})^T (i = b,w,t,l,r)$ denotes hydrodynamic force and moment on part i. $F_{bi} = (f_{bi}, \tau_{bi})^T (i = w,l,r)$ indicates the external force of body on part i. On the contrary, $F_{ib} = (f_{ib}, \tau_{ib})^T$ expresses the external force of part i on body, and the same explanation for F_{wt} or F_{tw}. The purpose of multiplying both sides of the equation by ${}^b H_i$ is to transform the forces and moments from joints frame to body frame. Thereby, it follows that ${}^b H_i F_{bi} + F_{ib} = 0$ since they are the interaction forces.

For the convenience of analysis and presentation, some parameters are defined: $\xi_t = {}^t \dot{H}_w V_w$ and $\xi_i = {}^i \dot{H}_b V_b \ (i = w,l,r)$.

Afterward, Equation 3.4 formalizes the kinematic of body and movable surfaces, so we can derive the basic form of each speed derivative.

$$
\begin{aligned}
\dot{V}_i &= {}^i H_b \dot{V}_b + {}^i \dot{H}_b V_b + \dot{\delta}_i \qquad (i = w,l,r) \\
\dot{V}_t &= {}^t H_w \dot{V}_w + {}^t \dot{H}_w V_w + \dot{\delta}_t,
\end{aligned}
\tag{3.24}
$$

where

$$
{}^i \dot{H}_b = \begin{pmatrix} -\dot{\theta}_i \hat{j}_i^i R_b & \dot{\theta}_i \hat{j}_i^i R_b^b \hat{P}_i \\ 0_{3\times 3} & -\dot{\theta}_i \hat{j}_i^i R_b \end{pmatrix} \qquad (i = w,l,r)
$$

and ${}^t H_b = {}^t H_w {}^w H_b$. Substituting Equation 3.24 into Equation 3.23 yields

$$
\sum_{i=b,w,t,l,r} {}^b H_i M_i \dot{V}_i = \left(\sum_{i=b,w,t,l,r} {}^b H_i M_i {}^i H_b \right) \dot{V}_b + \Pi_e,
\tag{3.25}
$$

where

$$\Pi_e = \sum_{i=b,w,t,l,r} {}^b H_i M_i (\xi_i + \dot{\delta}_i) + {}^b H_t M_t^{t} H_w (\xi_w + \dot{\delta}_w).$$

Similarly, the sum of all forces and moments by adding the right sides of Equation 3.23 can be derived. Furthermore, according to the equality of the left and right sides of the equation, the final kinetic equation can be taken in the following form:

$$M\dot{V}_b = -\Pi_e + \Pi_c + \Pi_h + \Pi_g + \Gamma_m + \Gamma_s, \tag{3.26}$$

where

$$M = \sum_{i=b,w,t,l,r} {}^b H_i M_i^{i} H_b$$

$$\Pi_c = -\sum_{i=b,w,t,l,r} {}^b H_i F_{ci} = -\sum {}^b H_i \begin{pmatrix} \hat{\Omega}_i & 0_{3\times3} \\ \hat{V}_i & \hat{\Omega}_i \end{pmatrix} M_i \begin{pmatrix} V_i \\ \Omega_i \end{pmatrix}$$

$$\Pi_h = \sum_{i=b,w,t,l,r} {}^b H_i F_{hi}$$

$$\Pi_g = G_b$$

$$\Gamma_m = m_m \begin{pmatrix} 2\hat{\dot{P}}_m \Omega_b - \ddot{P}_m \\ \hat{P}_m (2\hat{\dot{P}}_m \Omega_b - \ddot{P}_m) \end{pmatrix}$$

$$\Gamma_s = m_s \begin{pmatrix} 2\hat{\dot{P}}_s \Omega_b - \ddot{P}_s \\ \hat{P}_s (2\hat{\dot{P}}_s \Omega_b - \ddot{P}_s) \end{pmatrix}.$$

3.4 ANALYSIS OF THE STEADY GLIDING MOTION

In this section, a detailed hydrodynamic analysis of the dolphin-like glider during a steady gliding motion is presented. Note that the dolphin-like glider has controllable pectoral fins and flattened fluke that could change the hydrodynamic performance by adjusting their turning angles. Therefore, hydrodynamic forces on the dolphin body, pectoral fins, and the flattened fluke should be separately analyzed, which is different from the traditional gliders.

Figure 3.3 defines the coordinate frames including an inertial frame and a body reference frame to describe the gliding motion. In the inertial

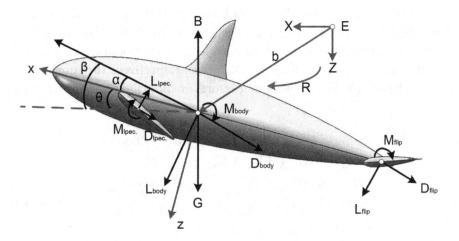

FIGURE 3.3   Coordinate systems defined to describe the steady gliding motion.

frame E_{xyz}, the horizontal axes of x and y are perpendicular to the gravity, and the z-axis is along the positive gravity direction. Note that the inertial frame is considered to be earth-fixed and no-rotating. For the body reference frame O_{xyz}, the coordinate center is fixed in the center of buoyancy of the glider, the x-axis is along the longitudinal axis of the dolphin-like glider from fluke to head, the y-axis is along the pectoral shafts from left to right, and the z-axis follows the right-hand rule. Let R and b separately denote the rotation matrix from the body frame to the inertial frame and the vector from the origin of the inertial frame to the origin of the body frame.

In order to obtain the hydrodynamic performance in a steady gliding motion, we first suppose the dolphin-like glider gets the hydrodynamic equilibrium in the vertical plane. Therefore, it is easy to get the vertical plane equilibrium equations as follows:

$$
\begin{aligned}
\dot{x} &= v_x \cos\theta + v_z \sin\theta \\
\dot{z} &= -v_x \sin\theta + v_z \cos\theta \\
0 &= (m_z - m_x)v_x v_z - m_b g(r_{b_x} \cos\theta + r_{b_x} \sin\theta) + M_{DL_{Total}}, \\
0 &= -m_0 g \sin\theta + L_{Total_x} + D_{Total_x} \\
0 &= m_0 g \sin\theta + L_{Total_z} + D_{Total_z}
\end{aligned} \qquad (3.27)
$$

where v_x and v_z are, respectively, the components of the gliding velocity in the x and z directions. θ denotes the pitch angle of the robotic dolphin.

m_0 and m_b are, respectively, the net buoyancy of the glider and the variable ballast point mass, which is offset r_b from the center of buoyancy. m_x and m_z are, respectively, the added mass terms corresponding to the x and z directions. $L_{Total_x}, L_{Total_z}, D_{Total_x}, D_{Total_z}, M_{DL_{Total}}$ are, respectively, the sum of the hydrodynamic forces on the body, pectoral fins, and fluke, as shown in the following equations:

$$L_{Total_x} = (L_{body} + L_{lpec} + L_{rpec} + L_{fluke})\sin\alpha_{body}$$
$$L_{Total_z} = -(L_{body} + L_{lpec} + L_{rpec} + L_{fluke})\cos\alpha_{body}$$
$$D_{Total_x} = -(D_{body} + D_{lpec} + D_{rpec} + D_{fluke})\cos\alpha_{body}$$
$$D_{Total_z} = -(D_{body} + D_{lpec} + D_{rpec} + D_{fluke})\sin\alpha_{body} \quad, \qquad (3.28)$$
$$L_i = 0.5\rho C_{L_i}(\alpha_i)S_i v^2$$
$$D_i = 0.5\rho C_{D_i}(\alpha_i)S_i v^2$$

where i = *body, lpec, repc,* and *fluke* denotes the related variable about the dolphin body, left pectoral fin, right pectoral fin, and flattened fluke, respectively. L_i, D_i respectively denote the hydrodynamic lift and drag on the body, left pectoral fin, right pectoral fin, and flattened fluke. α_i indicates the angle of attack of the glider. α_i (i = *lpec, repc, and fluke*) can be obtained through α_{body} and turn angles of pectoral fins and fluke. s_i denotes the maximum cross sectional area. ρ indicates the fluid density. v denotes the relative velocity of the robotic dolphin w.r.t. the fluid.

$$\theta = \arctan\left(-\frac{\sum C_{L_i}(\alpha_i)\sin\alpha_{body} - \sum C_{D_i}(\alpha_i)\cos\alpha_{body}}{\sum C_{L_i}(\alpha_i)\cos\alpha_{body} - \sum C_{D_i}(\alpha_i)\sin\alpha_{body}}\right) \quad (3.29)$$

According to the equilibrium equations, we easily obtain the relationship between pitch angle θ and the angle of attack α_i for the dolphin-like glider. We can find that the pitch angle θ only depends on the angle of attack α_i and the lift coefficient and the drag coefficient of the dolphin body, pectoral fins, and flatten fluke and is independent of the others. Therefore, we can easily control the pitch angle θ or the angle of attack α of the gliding underwater robot for an expected gliding motion through adjusting the turn angle β_i of pectoral fins and flatten fluke. Figure 3.4 depicts the pitch angle θ varying with the angle of attack of the body α_i with different turn angle β_i. According to the black and green curves in Figure 3.4, we can see that a little adjustment for β_{pec}, about 5°, even with $\beta_{fluke} = 0$°, could

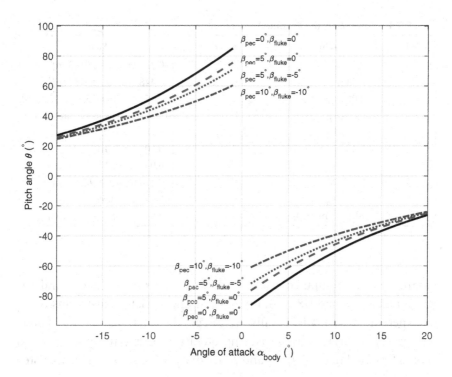

FIGURE 3.4 The relationship between pitch angle and the angle of attack.

successfully lead to an expected pitch angle θ. If we also adjust β_{fluke} at the same time, more apparent effects can be achieved, see the red and blue curves in Figure 3.4. These results illustrate that the controlled pectoral fins and flattened fluke will bring obvious effects for adjusting the gliding attitude.

3.5 RESULTS AND ANALYSES

In this chapter, we analyze the steady-state spiraling motion based on two motions which include the flapping and gliding motions. By simulation, we can draw some conclusions about the two motions. Subsequently, extensive aquatic experiments were carried out to verify the built dynamic model in a 2.4-m-deep diving pool.

3.5.1 Simulation Results

MATLAB®/Simulink is used to implement the simulations to verify the validity of the complete dynamic model. First, some physical parameters of the dynamic model are tabulated in Table 3.1. Furthermore, hydrodynamic

TABLE 3.1   Parameters of Dynamic Model

Symbol	Value	Symbol	Value
m_b	5.28 kg	^bP_w	$[-0.2257 \text{ m}, 0, -0.0123 \text{ m}]^T$
m_w	0.5 kg	^bP_t	$[-0.4488 \text{ m}, 0, -0.0123 \text{ m}]^T$
m_t	0.077 kg	^bP_l	$[0.1063 \text{ m}, -0.0615 \text{ m}, 0.0318 \text{ m}]^T$
m_l	0.074 kg	^bP_r	$[0.1063 \text{ m}, 0.0615 \text{ m}, 0.0318 \text{ m}]^T$
m_r	0.074 kg	P_b	$[0.001 \text{ m}, 0, 0.052 \text{ m}]^T$
m_m	0.42 kg	P_w	$[-0.109 \text{ m}, 0, 0]^T$
m_s	0.035 kg	P_t	$[-0.041 \text{ m}, 0, 0]^T$
g	9.8 m/s²	P_l	$[-0.014 \text{ m}, -0.05923 \text{ m}, 0]^T$
ρ	998.2 kg/m³	P_r	$[-0.014 \text{ m}, 0.05923 \text{ m}, 0]^T$
P_{j0}	$[0.1953 \text{ m}, 0, 0]^T$	P_{m0}	$[-0.04077 \text{ m}, 0, -0.03715 \text{ m}]^T$

parameters are computed using both CFD software packages and the curve-fitting method [184], and then some appropriate adjustments based on experiences are made. Finally, we can get the rotational inertia of body and movable surfaces by measuring quality property in SolidWorks. It should be noted that J_w, J_t, J_l, and J_r are diagonal matrix, which are listed in Table 3.1.

In the dolphin-like motion, the maximum joint amplitudes of the waist and tail and the phase between them are set at 30°, 45°, and 35°, respectively, and the frequency is 1 Hz. Simultaneously, the offset angles of left flipper and right flipper are 45° and 0. Figure 3.5 denotes 3-D trajectory within 50 s. We could see the robot only dives to the depth of 5.7 m, and its turning radius is nearly 3 m.

In the gliding motion, the offset angle of the left flipper and the right flipper are the same as the dolphin-like motion. Figure 3.6 illustrates 3-D trajectory within 50 s. The diving depth and turning radius in the motion are 8.7 m and less than 1 m.

In terms of spiraling movement, gliding mode, or flapping mode, asymmetric hydrodynamic that would produce steering yaw moment can be generated in the left and right sides of the body through the difference of flipper fins. The following rolling motion causes the lift and drag forces in the vertical plane, which makes the body roll. On the other hand, the horizontal components of the two forces generate a centripetal force for the steady-state spiraling movement. Thus, the projection of the steady 3-D trajectory of Figure 3.6 on the horizontal plane is circular.

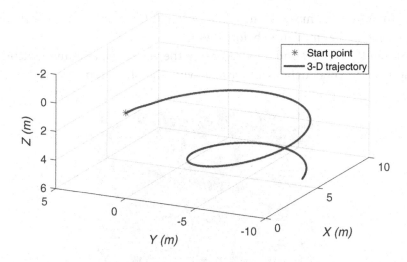

FIGURE 3.5   3-D trajectory of spiraling movements in dolphin-like motion.

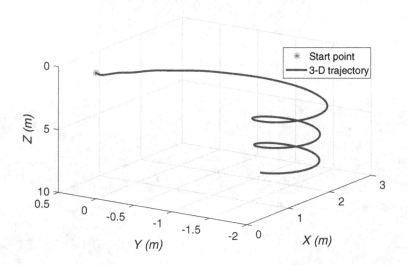

FIGURE 3.6   3-D trajectory of spiraling movements in the gliding motion.

3.5.2 Experimental Results

Experiments are implemented to validate the spiraling motion of the gliding robot in a 2.4 m depth diving pool. The snapshot sequences of the spiraling motion in gliding motion and dolphin-like motion are shown in Figures 3.7 and 3.8. Besides, in the gliding motion, due to the limited depth, we set the movable mass and piston at target positions, which are the same as the simulation before diving.

From Figures 3.9 and 3.10, regarding the yaw angle and diving depth, the experimental data is consistent with the simulation shape, which

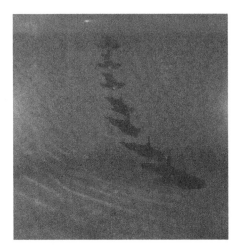

FIGURE 3.7   Snapshot sequence in the gliding motion.

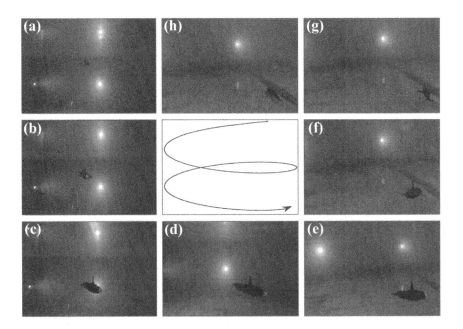

FIGURE 3.8   (a)–(h) Snapshot sequence in the dolphin-like motion.

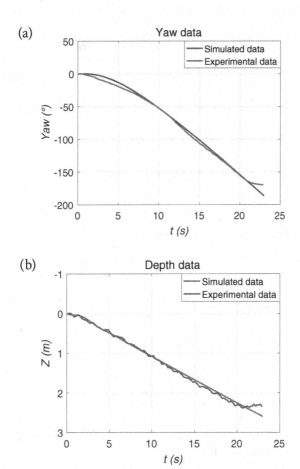

FIGURE 3.9 The simulated and experimental results in the dolphin-like motion. (a) Yaw angle. (b) Diving depth.

signifies that the 3-D completed dynamic model is valid. Furthermore, some conclusions can be drawn by comparing the spiraling movements of the gliding and dolphin-like motions. First, the diving depth in the gliding motion is greater than the dolphin-like motion within the same time, for both the flipper fins and the movable mass can supply the pitch moment in the spiraling movement, the latter of which brings about a higher effect. Moreover, since the net buoyancy effect has a certain delay, the diving speed in dolphin-like motion is faster than gliding in the early diving state, which directly changes the yaw angle in the dolphin-like motion, which is greater than that of the gliding motion. However, when the body posture

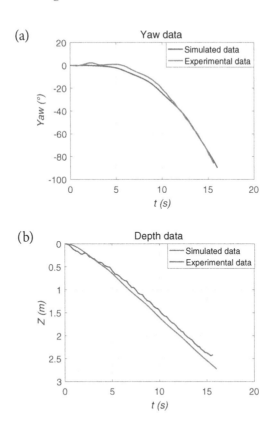

FIGURE 3.10 The simulated and experimental results in the gliding motion. (a) Yaw angle. (b) Diving depth.

in the gliding motion comes to a stable state, the gliding underwater robot could achieve a faster diving speed, which could offer a smaller turning radius. On the contrary, due to the dorsoventral propulsive mechanism, the robot has a relatively stable forward speed, which results in a stable change in the yaw angle. However, since the yaw moment in flapping is smaller than gliding motion in the late stage, the turning radius is much bigger.

3.6 CONCLUDING REMARKS

In this chapter, we proposed a complete dynamic model with full consideration of both 3-D gliding motion and dolphin-like motion for the gliding underwater robot and provided a detailed derivation process. By simulations based on the complete dynamic model, the difference in the spiraling movements of the dolphin-like motion and the gliding motion was compared. Via aquatic experiments, we analyzed the characteristics of spiraling movement, including the yaw angle and the diving depth, further proving that the dynamic model is effective. In contrast, we can conclude that the spiraling movement in the gliding motion is more stable and energy efficient than that in the dolphin-like motion.

REFERENCE

[184] J. Yu, J. Yuan, Z. Wu, and M. Tan, "Data-driven dynamic modeling for a swimming robotic fish," *IEEE Trans. Ind. Electron.*, vol. 63, no. 9, pp. 5632–5640, 2016.

CHAPTER 4

Depth Control of the Gliding Underwater Robot with Multiple Modes

4.1 INTRODUCTION

Depth control is a very important research topic for aquatic robots. For example, when autonomous underwater vehicles (AUVs) need to arrive at an area at a specified depth for more complex ocean tasks, it is crucial to achieve steady depth control. Besides, depth control is also an important part of navigation and path following. There exists much depth-control research for underwater robots. For the traditional AUVs, the propellers equipped in the vertical plane are usually used to generate the forces to dive or surface, which yet may cause environmental damage due to the noise. Li *et al.* presented an adaptive nonlinear controller by removing the assumption of a small pitch angle and realized the depth control for an AUV [185]. Silvestre *et al.* designed a nonlinear gain scheduling controller and applied the methodology to achieve depth control [186]. Wu *et al.* employed reinforcement learning for successful depth control [187]. Furthermore, for traditional underwater gliders, the depth control could be realized by adjusting the buoyancy-driven mechanism [188]. Nevertheless, the depth control for underwater gliders was studied

DOI: 10.1201/9781003347439-4

by simulations more than experiments. Regarding bio-inspired aquatic robots, the pitch moments were usually controlled to achieve further depth control via movable surfaces or a centroid adjustment mechanism. By regulating the movable pectoral surfaces, Yu *et al.* realized the depth control with the sliding mode fuzzy method on the prototypes of a robotic fish [189] and a robotic dolphin [190], of which the latter achieved a smaller steady error, less than 0.5 cm. Besides, Shen *et al.* employed a movable slider to adjust the centroid and further achieved depth control with the 5-cm error [191]. Moreover, Makrodimitris *et al.* conducted depth control with a pump mechanism equipped in a small robotic fish and realized a 2-cm steady error [136].

From the perspective of control methods, there are many model-based control methods that have been successfully applied for AUVs, such as sliding mode control (SMC), backstepping control, and model predictive control (MPC). In recent years, MPC has been widely used in mobile robots due to its good performance and less model dependence [192]–[194]. Generally, MPC is much suitable for underwater robot control since it can contribute to reducing the impact of large delays. However, there were relatively fewer applications of MPC on underwater robots due to the relatively complex implementation [195]–[197], and most of them were just designed in simulations. Therefore, more depth-control research on biomimetic aquatic robots with gliding motion, such as gliding underwater robots and gliding robotic fish, is necessary. However, depth control by only a buoyancy-driven mechanism is quite challenging due to large delay and insufficient control ability. More important, on one hand, the tasks which require to arrive at a target depth can be accomplished by the combination of gliding and dolphin-like motions. On the other hand, successful depth control by only a buoyancy-driven mechanism can provide the basis for the gliding underwater robot to achieve hovering, which is a significant technology for underwater operation.

Further progress is made in this chapter, whose objective is to achieve depth control with both gliding and dolphin-like motions for an underwater gliding robot. The contributions made in this chapter are summarized as follows:

- In terms of gliding motion, a novel depth control strategy consisting of an MPC depth controller, a velocity-proportional-integral-derivative (PID) heading controller, and a sliding mode observer (SMO) is proposed for the gliding motion.

- For dolphin-like motion, a control framework that achieves depth control by combining line-of-sight (LOS) with an adaptive control approach (ACA) is proposed. Specifically, we optimize the controller parameters based on extensive offline simulations of a full-state dynamic model.

- Extensive simulations and aquatic experiments are carried out. The obtained results validate the effectiveness of the proposed methods.

The remainder of this chapter is organized as follows. Section 4.2 provides the gliding motion depth control, including the model simplification, heading controller, and SMO. Section 4.3 presents the dolphin-like motion depth control, including the dual-mode design and ACA. Section 4.4 shows the simulation and experimental results followed by a discussion. Section 4.5 concludes this chapter with an outline of future work.

4.2 DEPTH CONTROL IN GLIDING MOTION

4.2.1 Problem Statement

The depth control in the gliding motion mainly uses depth as the feedback to control the net buoyancy, thereby realizing vertical motion. However, there are mainly four problems as follows:

1. The displacement of the buoyancy adjustment mechanism is limited, and the net buoyancy adjustment range is small, resulting in a relatively insufficient buoyancy-driving capacity and prone to overshoot.

2. It is difficult to directly measure the velocity of underwater motion, so it is hard to provide effective state information for model-based motion control algorithms.

3. During the depth control process of the gliding motion, the yaw direction is prone to unnecessary deflection. If a complete dynamic model is used, it will not only increase the difficulty of the controller design but also consume embedded computing resources.

4. The buoyancy adjustment mechanism changes the volume by moving the injector at a high frequency, which may cause system instability.

Aiming at the above problems, this chapter proposes many solutions. For problem (1), a depth controller based on a model prediction method is designed that does not require high model accuracy and is suitable for underwater robot motion control. By predicting the output of the dynamic model, the control quantity can be calculated in real time, and the target diving speed is designed based on the Bezier curve, thereby reducing the overshoot. For problem (2), the SMO is used to estimate the motion speed, and the real diving speed returned by the depth sensor is combined with the attitude information to estimate the gliding speed. For problem (3), the dynamic model is simplified by ignoring the yaw and roll motions. In order to satisfy this assumption, a yaw controller is introduced to ensure that the yaw angle does not have a large deflection. For problem (4), the injector is driven by a motor. If it is in position mode and control commands are issued frequently, the motor will often be in a state of acceleration and deceleration. Therefore, we put the motor in speed mode and design an internal position loop based on a PID controller to smooth the acceleration and deceleration process. Based on the preceding ideas, Figure 4.1 shows the block diagram of the gliding modal depth control system.

4.2.2 Simplified Plant Model

This section presents a depth control system to achieve the target depth in the gliding motion. Taking the calculation efficiency of an embedded platform into consideration, a simplified model is applied for the model predictive control method. In this work, we only consider the dynamic model of the vertical plane. Therefore, we assume that the motions of yaw and roll are negligible in this model. Hence, the kinematic model is as follows:

$$\dot{x} = u\cos(\theta) + w\sin(\theta)$$
$$\dot{z} = -u\sin(\theta) + w\cos(\theta),$$
$$\dot{\theta} = q$$

(4.1)

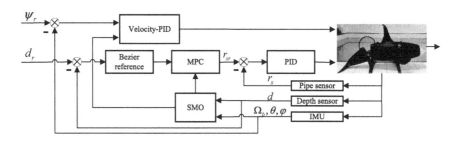

FIGURE 4.1   Framework of the depth control system in the gliding motion.

where x and z denote the inertial coordinate position in the horizontal and vertical directions, respectively. θ is pitch angle. Denote u, v, and w be the velocity with respect to (w.r.t.) body frame. Correspondingly, the dynamic equation is as follows:

$$M\dot{v} = -C(v)v - Dv + \tau, \qquad (4.2)$$

where $v = [u, w, q]^{T}$ denotes the forward velocity, diving velocity, and angular velocity about the y-axis w.r.t. body frame, respectively. $M = diag\{m_{1}, m_{2}, m_{3}\}$ represents the rigid-body inertia, including added mass. $D = diag\{d_{1}, d_{2}, d_{3}\}$ is the damping matrix which is simplified to a constant matrix. $\tau = (\tau_{u}, \tau_{w}, \tau_{q})^{T}$ illustrates the net buoyancy vector. As for the $C(v)$, it is Coriolis and centripetal matrix, which can be defined as follows:

$$C(v) = \begin{pmatrix} 0 & 0 & m_{2}w \\ 0 & 0 & -m_{1}u \\ -m_{2}w & m_{1}u & 0 \end{pmatrix}. \qquad (4.3)$$

Therefore, based on the above analysis, we can obtain the following equation:

$$\ddot{z} = -\frac{d_{1}}{m_{1}}\dot{z} + \left(\frac{\tau_{w}\cos(\theta)}{m_{2}} - \frac{\tau_{u}\sin(\theta)}{m_{1}} \right) + L, \qquad (4.4)$$

where

$$L = \left(\frac{d_{1}}{m_{1}} - \frac{d_{2}}{m_{2}} \right) w\cos(\theta) + \left(\frac{m_{1} - m_{2}}{m_{2}} \right) uq\cos(\theta) + \left(\frac{m_{2} - m_{1}}{m_{1}} \right) wq\sin(\theta).$$

It can be seen from the above formula that the vertical dynamic model of the gliding underwater robot is cumbersome, and the nonlinear term L makes it difficult to design the control system. In addition, in this section, a model prediction algorithm is adopted to design a depth controller whose performance depends largely on the prediction period and the control period. However, the matrix formed by L needs to be calculated in real time, which is basically impossible to implement in the embedded system. Furthermore, the buoyancy adjustment mechanism is applied without moving the slider when controlling the depth.

Through observation, it is found that during the small-range movement of the buoyancy adjustment mechanism, the gliding attitude of the body is basically parallel to the horizontal plane; that is, the pitch angle is basically less than 10°, and even close to 0° when approaching the target depth. Therefore, in order to save the computing resources of the embedded system, it is assumed that $\sin(\theta) \approx 0$, and $\cos(\theta) \approx 1$ in the controller design process, that is, converted to a linear model, but the state update process still uses the original model. Therefore, when constructing the control law, the control variable in the vertical direction can be expressed as the net buoyancy u_c. In addition, considering that the motion control period of a typical glider is about 2 s to 10 s, we set the sampling time to 1 s, by converting the simplified the dynamic model by discretizing the expression, we can get

$$w(k+1) = Aw(k) + Bu_c(k) + L(k), \tag{4.5}$$

where

$$A = 1 - \frac{d_2}{m_2}$$

$$B = \frac{1}{m_2} \qquad .$$

$$L = \frac{m_1}{m_2} u(k)q(k)$$

4.2.3 SMO and Heading Controller Design

According to the dynamic model, we should calculate the predictive diving speed in real time. Hence, u and q should also be attained. However, it may be inaccurate to obtain the velocity vector just by means of the simplified dynamic model. Therefore, an SMO is applied to decrease the estimated error. The diving speed and attitude information can be measured by depth sensor and attitude sensor, respectively. Based on the obtained sensor data, we can estimate the forward speed.

Firstly, the real diving speed w.r.t inertial frame can be calculated by the depth sensor. Denoting $^g\tilde{U}_{bz}$ as the estimated velocity, the estimation error can be defined as s.

$$^gU_{bz} = \dot{d}$$

$$s = {}^gU_{bz} - {}^g\tilde{U}_{bz} \tag{4.6}$$

Based on the completed kinematics and estimation error, the SMO can be designed as follows:

$$
{}^g\dot{\tilde{U}}_b = {}^gR_b\check{\Omega}_b\tilde{U}_b + {}^gR_b\dot{\check{U}}_b + \begin{pmatrix} c_x \\ c_y \\ c_z \end{pmatrix}\mathrm{sat}(s), \tag{4.7}
$$

where

$$
\dot{\check{U}}_b = M^{-1}\left(-C\left(\tilde{U}_b\right)\tilde{U}_b - D\tilde{U}_b + \tau\right)
$$

(c_x, c_y, c_z) is the weight vector of SMO. By setting suitable parameters, a Hurwitz matrix can be made to guarantee the convergence of forward speed. Besides, a saturation function $\mathrm{sat}(s)$ is applied to alleviate the undesirable chattering effect. Thereafter, the heading controller can be designed based on the obtained forward speed.

Regarding the heading control, it should be noted that the control signal refers to the offset angles of flippers. By deflecting the flippers, differential steering moments can be generated further to produce the steering forces and moments. Due to the relatively poor maneuverability for gliding motion, the heading control process may be longer and easy to overshoot. Since the steering forces and moments are closely related to speed, we apply a heading controller, which exerts larger control signal when the speed of robot is low and smaller control signal when the speed is high. This design can decrease the overshoot to some extent. By setting the maximum gliding speed of the gliding underwater robot v_{max}, a weight coefficient k_f can be calculated as follows:

$$
k_f = \frac{\sqrt{v_x^2 + v_z^2}}{v_{max}}, \tag{4.8}
$$

where v_{max} denotes the max gliding velocity. Furthermore, based on the yaw error e_ψ, a PID controller is employed to obtain the final control signal as follows:

$$
u_f = k_f\left(k_p e_\psi + k_i \int e_\psi + k_d \dot{e}_\psi\right). \tag{4.9}
$$

By adjusting the PID parameters, the control variable can be directly mapped to the flipper angles $\kappa = [\kappa_l, \kappa_r]$. It is worth noting that only unilateral flippers are used for yaw adjustment in this section, that is, when one flipper is deflected, the other flipper angle is 0.

$$\begin{cases} \kappa_l = u_f, \kappa_r = 0 & e_\psi < 0 \\ \kappa_l = 0, \kappa_r = u_f & e_\psi \geq 0 \end{cases} \tag{4.10}$$

4.2.4 Depth Controller Design

Based on the diving dynamic, we apply the model predictive methodology to design a depth controller. For a better explanation of the MPC method, we define the state variable as follows:

$$\xi(k|t) = \begin{pmatrix} w(k|t) \\ u_c(k-1|t) \end{pmatrix}, \tag{4.11}$$

where $(k|t)$ represents the predicted value of the future k time based on the time t. Afterward, the dynamic can be formalized as

$$\begin{aligned} \xi(k+1|t) &= \tilde{A}\xi(k|t) + \tilde{B}\Delta u_c(k|t) + \tilde{L}(t) \\ \eta(k|t) &= \tilde{C}\xi(k|t) \end{aligned}, \tag{4.12}$$

where

$$\tilde{A} = \begin{pmatrix} A & B \\ 0 & 1 \end{pmatrix}; \tilde{B} = \begin{pmatrix} B \\ 1 \end{pmatrix}; \tilde{C} = (1 \ \ 0); \tilde{L} = \begin{pmatrix} L \\ 0 \end{pmatrix}.$$

Moreover, we denote the N_c and N_p as the control and predictive steps, respectively. Hence, the future states can be derived by iteration calculation.

$$\begin{aligned} \xi(k+2|t) &= \tilde{A}^2\xi(k|t) + \tilde{A}\tilde{B}\Delta u_c(k|t) + \tilde{B}\Delta u_c(k+1|t) \\ &\quad + (\tilde{A}+I)\tilde{L}(t) \end{aligned}$$

$$\vdots$$

$$\begin{aligned} \xi(k+N_c|t) &= \tilde{A}^{N_c}\xi(k|t) + \tilde{A}^{N_c-1}\tilde{B}\Delta u_c(k|t) + \cdots \\ &\quad + \tilde{B}\Delta u_c(k+N_c-1|t) + \left(\sum_{i=0}^{N_c-1}\tilde{A}\right)\tilde{L}(t) \end{aligned} \tag{4.13}$$

$$\vdots$$

$$\begin{aligned} \xi(k+N_p|t) &= \tilde{A}^{N_p}\xi(k|t) + \tilde{A}^{N_p-1}\tilde{B}\Delta u_c(k|t) + \cdots \\ &\quad + \tilde{A}^{N_p-N_c-1}\tilde{B}\Delta u_c(k+N_c|t) + \left(\sum_{i=0}^{N_p-1}\tilde{A}\right)\tilde{L}(t) \end{aligned}$$

Specially, we assume that L item keeps a constant in one predictive process to decouple the diving speed with forward speed and pitch angular speed. Move one step further, we can obtain the final form

$$Y(t) = Y\xi(k|t) + H\Delta U_c(t) + \Delta, \tag{4.14}$$

where

$$Y(t) = \left(\eta(k+1|t), \cdots, \eta(k+N_c|t), \cdots, \eta(k+N_p|t)\right)^T$$

$$Y = (\tilde{C}\tilde{A}, \cdots, \tilde{C}\tilde{A}^{N_c}, \cdots, \tilde{C}\tilde{A}^{N_p})^T$$

$$\Delta = \left(I, \cdots, \sum_{i=0}^{N_c-1}\tilde{A}, \cdots, \sum_{i=0}^{N_p-1}\tilde{A}\right)^T \tilde{L}(t)$$

$$H = \begin{pmatrix} \tilde{C}\tilde{B} & 0 & \cdots & & 0 \\ \vdots & & \cdots & \ddots & \vdots \\ \tilde{C}\tilde{A}^{N_c-1}\tilde{B} & \cdots & \cdots & & \tilde{C}\tilde{B} \\ \vdots & & \cdots & \cdots & \vdots \\ \tilde{C}\tilde{A}^{N_p-1}\tilde{B} & \cdots & \cdots & & \tilde{C}\tilde{A}^{N_p-N_c-1}\tilde{B} \end{pmatrix}$$

Afterward, the optimal solution of the control signal is calculated by optimizing an objective function for MPC method. The selection of this function should be considered via two factors. First, the steady error, the difference between the target depth and achieved depth, should be controlled to minimum. In this work, we transform the controlled the target from depth to diving speed. By designing a suitable reference trajectory of diving speed, the target depth can be arrived at. Second, the control increment cannot be drastic; otherwise, it may cause mechanical and electrical damage to the robot. Based on the preceding considerations, we adopt the objective function as follows:

$$J(\xi(t), \Delta U_c(t)) = \sum_{i=1}^{N_p} \left\| \eta(t+i|t) - \eta_{ref}(t+i|t) \right\|_Q^2 + \sum_{i=0}^{N_c-1} \left\| \Delta u_c(t+i|t) \right\|_R^2, \tag{4.15}$$

where Q and R are weight matrices. The first part guarantees the system's ability to follow the reference trajectory. The second one reflects the requirement for a smooth change in the control signal. By setting $E(t) = Y\xi(k|t) + \Delta - Y_{ref}(t)$, we can get the initial quadratic form of objective function as follows:

$$J\left(\xi(t), \Delta U_c(t)\right) = \Delta U_c(t)^T \Omega \Delta U_c(t) + \Psi \Delta U_c(t) + \Phi, \qquad (4.16)$$

where

$$\Omega = H^T \tilde{Q} H + \tilde{R}$$
$$\Psi = 2E(t)^T \tilde{Q} H$$
$$\Phi = E(t)^T \tilde{Q} E(t)$$
$$\tilde{Q} = diag\{Q, \cdots, Q\}_{N_x \times N_p}$$
$$\tilde{R} = diag\{R, \cdots, R\}_{N_u \times N_c}).$$

It is obviously that Φ is a constant in one optimization process due to the constant matrix $\tilde{Q}$. The N_x and N_u represent the dimension of state and control variables, N_u respectively. In this work, the state variable is diving speed, and the control variable is net buoyancy. Therefore, on a basis of the preceding analysis and derivation, we solve the depth control as an optimization issue. The final optimization function and constraints are as follows:

$$\min_{\Delta U_c(t)} J\left(\xi(t), \Delta U_c(t)\right) = \Delta U_c(t)^T \Omega \Delta U_c(t) + \Psi \Delta U_c(t)$$
$$s.t. \begin{cases} \Delta U_c(t) \in [\Delta U_{c\min}, \Delta U_{c\max}] \\ U_c(t) \in [U_{c\min}, U_{c\max}] \end{cases} \qquad (4.17)$$

By calculating the optimal solution, the optimal control increment sequence can be obtained:

$$\Delta U_c^* = \left(\Delta u_c(t)^* \quad \cdots \quad \Delta u_c(t + N_c - 1)^*\right)^T. \qquad (4.18)$$

Moving one step further, we can attain the final control signal by selecting its first item.

$$u_c(t) = u_c(t-1) + \Delta u_c(t)^* \qquad (4.19)$$

Since the robot adjusts the net buoyancy by moving the piston or pipe, we should map the control signal to the actual motor execution via calculating the net buoyancy.

$$r_{sr} = \frac{u_c(t)}{\rho g S},$$ (4.20)

where r_{sr} illustrates the reference pipe position in real time, ρ denotes the water density, g elucidates gravity acceleration. And S represents the bottom area of the pipe.

More important, the reference trajectory of diving speed is another key design in depth control system. For large time-delay systems, a good reference trajectory can be designed to achieve early action and avoid excessive overshoot. Hence, we design the reference trajectory based on the Bezier curve. As we all know, Bezier curves are commonly used to smooth the path. By setting different control points, the curve shapes are changed. Moreover, Bezier curves are also related to time domain. In our depth system, we hope the depth curve can reduce overshoot as much as possible. Therefore, based on the predictive steps, we apply quadratic Bezier curves to the reference trajectory of depth first. Afterward, the dot of depth reference can be calculated to offer the real-time diving speed reference trajectory. The reference trajectory of depth is designed as follows:

$$P_{ref}(i) = \left(1 - t(i)\right)^2 d + 2t(i)\left(1 - t(i)\right)^2 d_r + t(i)^2 d_r,$$ (4.21)

where

$$t = \frac{1}{N_p}\left(0, \ 1, \ \cdots, \ N_p\right),$$

d_r denotes the target depth, and d represents the real-time depth. Thereafter, the reference trajectory of diving speed can be derived by $V_{ref} = \dot{P}_{ref}$. To further control the diving speed better, we employ a segmented reference trajectory. By setting the depth threshold $d_{threshold}$ and segmented parameters, the final reference trajectory of diving speed can be formalized by

$$Y_{ref} = \begin{cases} c_1 V_{ref} & if \ |d_r - d| > d_{threshold} \\ c_2 V_{ref} & otherwise \end{cases},$$ (4.22)

ALGORITHM 4.1   Algorithm for the Control System

1: Simplify the dynamic model and discrete it
2: Initialize the model and control parameters
3: **repeat**
4: Obtain the depth and attitude data from sensors
5: Calculate the estimated velocity by SMO of Equation 4.7
6: **repeat** Heading control module
7: Use the estimated velocity to get k_f in Equation 4.8
8: Apply the k_f of PID controller of Equation 4.9
9: **until** $\psi_e < \psi_{threshold}$
10: **repeat** Depth control module
11: Optimize the reference tracking by Equations 4.21 and 4.22
12: Calculate the N_p-steps model outputs by Equation 4.14
13: Obtain the $\Omega, \Psi, \Phi, \tilde{Q}, \tilde{R}$
14: Solve the N_c-steps control outputs by calculating the gradient
15: Select the first control increment as the final output
16: **until** $d_e < d_{threshold}$
17: **until** Received the end command

---

where c_1 and c_2 are weight parameters. In particular, it should be noted that only two segments are designed in this chapter. Actually, the numbers of segments can be set in accordance with different target depth. In general, there can be more segments with large target depths, or a continuous function can be applied to illustrate. In order to illustrate the control procedure more clearly, we present the algorithm flowchart as Algorithm 4.1.

4.3 DEPTH CONTROL IN GLIDING MOTION

4.3.1 Problem Statement

The depth control in the dolphin-like motion mainly uses the depth or the pitch angle as the feedback and relies on the propulsive system as the driving force. The movable surfaces are usually applied to generate the pitching moment to realize up and down motions. The depth control in this motion mainly has the following problems:

1. In the existing research on depth control of underwater robots, many studies use the depth as the direct feedback quantity to design the controller. However, in the process of depth determination, most of the control joints of the bionic robotic fish are used to generate the pitch

moment. When the actual depth does not reach the expected value, the controller will keep the output; that is, the pitch joint will continue to generate torque, which will cause the body posture to be easily deflected. When the robot is about to reach the target depth, it is prone to over-shoot. During the high-speed vertical motion of the robot, the motion direction is roughly similar to the direction of the pitch angle. Using the pitch angle as the control feedback will make the body posture smoother.

2. Most of the existing depth control methods of bionic robotic dolphins imitate the deflection of pectoral fins to generate pitching moment. However, for the dolphin-like motion, flippers are the only mechanism to achieve yaw motion and play a crucial role in three-dimensional motion. If the flippers are applied for yaw and pitch control at the same time, it will inevitably lead to a coupling phenomenon. In addi-tion, the pitching moment generated by the flipper deflection is closely related to the movement speed. When the forward speed of the body is small, a satisfactory pitching moment may not be generated, so it is more suitable for the depth maintenance stage.

3. In the depth-control process of the dolphin-like motion, due to the complex underwater environment, the highly nonlinear and strong coupling characteristics, the system has high requirements for con-trol parameters, making it difficult to obtain suitable parameters.

In response to preceding problems, we propose a dual-mode depth-control method by combining LOS with an adaptive control approach.

4.3.2 LOS Method

In this chapter, we aim to make the robot swim at the desired depth using dolphin-like motions. Generally, with sufficient propulsion and a suitable lift-to-drag ratio, such a robot can easily dive with a stable pitch angle. Thus, we can transform the control target from the depth to the pitch angle. Another benefit of controlling the pitch angle is that the track-ing path can be smoother, since the robot's body attitude cannot change sharply. In principle, depth control is a two-dimensional issue and can be simplified to identify tracking points at the desired depth. We can then employ LOS guidance to map the desired tracking points to pitch angles. As shown in Figure 4.2, given the target depth d, we consider a vector $\vec{a}$ that is perpendicular to z_g. Then, taking the robot's centroid as the center, we draw a circle with radius R that intersects the vector $\vec{a}$ at the points A

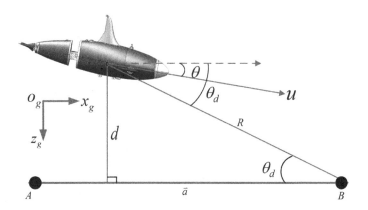

FIGURE 4.2   Diagram showing our LOS implementation.

and B. Then, we select the point B as the target point. Based on the real-time depth z, we can then obtain the target pitch angle as

$$\theta_d = \arctan(\frac{d_e}{\vec{a}}),\tag{4.23}$$

where

$$\begin{cases} d_e = d - z, \\ d_e^2 + \|\vec{a}\|^2 = R^2. \end{cases}$$

Here, $\|\cdot\|$ indicates the Euclidean norm.

4.3.3 Control Framework

In this chapter, we use a central pattern generator (CPG) model to control the robotic dolphin's swimming mode, as this can effectively ensure the control signals produce smooth motion transitions, even in the face of sudden parameter changes. This CPG model can be expressed as [198]

$$\begin{cases} \dot{\alpha}_i = 2\pi f_i + \sum_{j \in Y(i)} \omega_{ij} \sin\left(\alpha_j - \alpha_i - \phi_{ij}\right), \\ \ddot{r}_i = a_i \left(\frac{a_i}{4}\left(A_i - r_i\right) - \dot{r}_i\right), \\ x_i = r_i \left(1 + \cos\alpha_i,\right) \end{cases}\tag{4.24}$$

where r_i and α_i denote the amplitude and phase of the ith oscillator, respectively. In addition, ϕ_{ij} and ω_{ij} represent the phase difference and weight between the ith and the jth oscillator, respectively. a_i is a strictly positive constant. A_i and f_i indicate the intrinsic amplitude and frequency, respectively. Y_i denotes the set of oscillators from which the ith oscillator receives inbound couplings, and x_i represents the model's output.

Since the CPG model cannot change the offset of the output signals when we employ the one-sided CPG output to save the startup times, we should also add an offset angle β_i based on the output angle φ_i calculated by x_i.

Furthermore, in order to produce diving and surfacing motions, the robot needs to create suitable pitch moments. We therefore selected two control variables, namely, the flipper deflection angle φ and the CPG model offset β_i. By controlling these two signals, the robot can adjust its pitch moment to float up or down. Figure 4.3 illustrates the different control options and the corresponding forces.

Specifically, by changing the deflection angle ψ of the flippers, the water flow can result in upward or downward forces, which can in turn produce positive or negative pitch moments as of the center of gravity changes. The robot can also use the offset β_i to change the direction of the

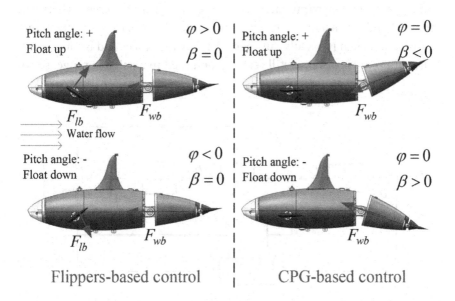

FIGURE 4.3   Illustration of the control signals for both modes.

propulsive forces from its body and flukes. Clearly, a nonzero β_i means F_{wb} is no longer horizontal but rather points obliquely upward or downward. Therefore, we combine the LOS and CPG control methods to create the proposed control algorithm framework, as illustrated in Figure 4.4. This enables the depth control target to be reached via an iterative process.

4.3.4 Controller Design

Since our control target is the pitch angle, we describe the dolphin's angular motion by the following model [199]:

$$\dot{\theta}_{k+1} = \begin{pmatrix} p_k & q_k \end{pmatrix} \begin{pmatrix} \dot{\theta}_k \\ u_k \end{pmatrix}, \tag{4.25}$$

where

$$\begin{cases} \begin{pmatrix} p_k \\ q_k \end{pmatrix} = \chi_k + \zeta_k = \begin{pmatrix} a_k \\ b_k \end{pmatrix} + \begin{pmatrix} \zeta_k^0 \\ \zeta_k^1 \end{pmatrix} = \begin{pmatrix} \hat{a}_k + \tilde{a}_k \\ \hat{b}_k + \tilde{b}_k \end{pmatrix} + \begin{pmatrix} \zeta_k^0 \\ \zeta_k^1 \end{pmatrix}. \\ \|\zeta_k\| \le \lambda \eta_k \end{cases}$$

Here, $\dot{\theta}_k$ represents the rate of pitch angle change, and u_k is the control variable that provides the pitch moment. Here, u_k is either ψ or β_i. Next, $\hat{a}_k$ and $\hat{b}_k$ are adaptive parameters, and $\tilde{a}_k$ and $\tilde{b}_k$ are the differences between the real and estimated values. In addition, ζ_k denotes the parameter perturbation, which is usually due to external disturbances and is thus generally random. However, it still remains bounded in a suitable range, so we

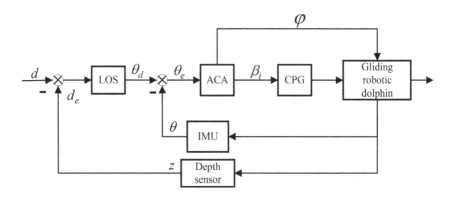

FIGURE 4.4   Framework of the depth control system in the dolphin-like motion.

assume it is less than $\lambda \eta_k$, where η_k is a positive scalar value that we will estimate and λ is a positive, manually adjusted parameter that controls the rate at which η_k changes. We also define q_{min} as the minimum value of q_k, and assume that $q_k > 0$ due to the sign of the variable u_k; that is, $q_k > q_{min} > 0$.

The pitch angle is related to its rate of change via

$$\dot{\theta}_{k+1} = \theta_{k+1} - \theta_k. \tag{4.26}$$

Substituting this into Equation 4.25 then gives us the final form:

$$\theta_{k+1} = (1 + p_k)\theta_k - p_k\theta_{k-1} + q_k u_k. \tag{4.27}$$

By setting the reference pitch angle $\theta_{k+1}^r = \theta_d$ and inverting Equation 4.27, we can express the control signal u_k as

$$u_k = \frac{\theta_{k+1}^r - (1 + \hat{a}_k)\theta_k + \hat{a}_k\theta_{k-1}}{\hat{b}_k} \tag{4.28}$$

In addition, due to structural limitations, u_k must be bounded, lying within a range $u_k \in [u_{min}, u_{max}]$ that is determined by the robot's design. Next, in order to develop the adaptation rules, we need to calculate the control error. The tracking error relative to the control target can be defined as

$$e_{k+1} = \theta_{k+1}^r - \theta_{k+1} = \theta_{k+1}^r - (1 + p_k)\theta_k + p_k\theta_{k-1} - q_k u_k + \hat{b}_k u_k - \hat{b}_k u_k, \tag{4.29}$$

where u_k is not in the saturation zone. Hence, we can substitute further obtain

$$\begin{aligned} e_{k+1} &= -(\tilde{a}_k + \zeta_k^0)(\theta_k - \theta_{k-1}) - \hat{b}_k^{-1}(\tilde{b}_k + \zeta_k^1)(\theta_{k+1}^r - (1 + \hat{a}_k)\theta_k + \hat{a}_k\theta_{k-1}) \\ &= -\left(\tilde{a}_k + \zeta_k^0 \quad \tilde{b}_k + \zeta_k^1\right)\delta_k \end{aligned}, \tag{4.30}$$

where

$$\delta_k = \begin{pmatrix} \dot{\theta}_k \\ u_k \end{pmatrix}.$$

Thus, we can express the expected control error magnitude as

$$|e_k| \le \|\zeta_k\| \|\delta_k\| \le \lambda \eta_k \|\delta_k\|. \tag{4.31}$$

4.3.5 Adaptation Rules

Based on the control error, we can now derive the adaptation rules. First, the parameter adaptation rate should decrease with the error, stopping entirely once the error satisfies $|e_k| \leq \lambda \eta_k \|\delta_k\|$. Since we are aiming to estimate the parameter η_k, let the estimated value be $\hat{\eta}_k$. In addition, we define a coefficient c_k that controls the adaptation rate as [199]

$$c_k = \begin{cases} (1 - \lambda \hat{\eta}_{k-1} \|\delta_{k-1}\| |e_k|^{-1}), \text{if} |e_k| \geq \lambda \hat{\eta}_{k-1} \|\delta_{k-1}\| \\ 0, \text{otherwise} \end{cases} \tag{4.32}$$

Note that c_k lies within the range $[0,1)$. Then, we can estimate η_k as

$$\hat{\eta}_k = \hat{\eta}_{k-1} + \sigma_k^{-1}(\lambda \gamma c_k |e_k| \|\delta_{k-1}\|), \tag{4.33}$$

where

$$\sigma_k = 1 + \delta_{k-1}^T \Lambda \delta_{k-1} + \gamma \lambda^2 \|\delta_{k-1}\|^2$$

$$\Lambda = \begin{pmatrix} \mu_1 & 0 \\ 0 & \mu_2 \end{pmatrix} .$$

Here, $|\cdot|$ denotes the absolute value. In addition, γ is a positive parameter to be set manually, and μ_1 and μ_2 are positive constants. In this chapter, we set $\mu_1 = \mu_2 = \mu$ and $\hat{\eta}_0 = 0$. Finally, we can obtain the adaptation rule for the adaptive model as

$$\begin{pmatrix} \hat{a}_k \\ \hat{b}_k \end{pmatrix} = \begin{pmatrix} \hat{a}_{k-1} \\ \hat{b}_{k-1} \end{pmatrix} - \frac{c_k e_k}{\sigma_k} \Delta \delta_{k-1}, \tag{4.34}$$

where $\hat{b}_k > b_{\min}$.

Proposition: If the ACA is employed for the preceding design, the estimated variable values are bounded as confirmed by Verma *et al.* [199].

Proof: First, define $\tilde{y}_k = \begin{pmatrix} \tilde{a}_k & \tilde{b}_k \end{pmatrix}$ and $\tilde{\eta}_k = \eta_k - \hat{\eta}_k$. Next, consider the nonnegative Lyapunov function

$$V_k = \tilde{y}_k^T \Lambda^{-1} \tilde{y}_k + \frac{1}{\gamma} \tilde{\eta}_k^2 . \tag{4.35}$$

Then, the change in this function can be expressed as

$$\Delta V_k = V_k - V_{k-1} = P_1 + P_2, \tag{4.36}$$

where

$$P_1 = \tilde{y}_k^T \Lambda^{-1} \tilde{y}_k - \tilde{y}_{k-1}^T \Lambda^{-1} \tilde{y}_{k-1}$$

$$P_2 = \frac{1}{\gamma}\left(\tilde{\eta}_k^2 - \tilde{\eta}_{k-1}^2\right).$$

For P_2, we can obtain $\tilde{\eta}_k$ as follows:

$$\tilde{\eta}_k = \tilde{\eta}_{k-1} - \sigma_k^{-1}(\lambda \gamma c_k |e_k| \|\delta_{k-1}\|) . \tag{4.37}$$

Thus, we can express P_2 as

$$P_2 = \frac{c_k^2 e_k^2 \gamma \lambda^2}{\sigma_k^2}\|\delta_{k-1}\|^2 - \frac{2 c_k \lambda}{\sigma_k}\|\delta_{k-1}\||e_k|\tilde{\eta}_{k-1}. \tag{4.38}$$

In addition, since $\tilde{b}_k > b_{\min}$, the adaptation rule can be reexpressed as

$$m_k = \hat{y}_{k-1} - \frac{c_k e_k}{\sigma_k}\Delta\delta_{k-1}, \tag{4.39}$$

where

$$\hat{y}_k = f(m_k) = \begin{cases} \left(m_k^0 \quad m_k^1\right) & \text{if}(m_k^1 > b_{\min}) \\ \left(m_k^0 \quad b_{\min}\right) & \text{if}(m_k^1 < b_{\min}) \end{cases}.$$

Thus, we can conclude that $\|y_{k-1} - f(m_k)\| \le \|y_{k-1} - m_k\|$. As a result, $\tilde{y}_k^T \Lambda^{-1} \tilde{y}_k \le (y_{k-1} - m_k)^T \Lambda^{-1}(y_{k-1} - m_k)$, and we can derive the following bound on P_1

$$P_1 \le \frac{2 c_k e_k}{\sigma_k} \tilde{y}_{k-1}^T \delta_{k-1} + \frac{c_k^2 e_k^2}{\sigma_k^2} \delta_{k-1}^T \Lambda \delta_{k-1}. \tag{4.40}$$

In addition, we can obtain the following bound on ΔV_k:

$$\Delta V_k \le \frac{2 c_k e_k}{\sigma_k} \tilde{y}_{k-1}^T \delta_{k-1} - \frac{2 c_k \lambda}{\sigma_k}\|\delta_{k-1}\||e_k|\tilde{\eta}_{k-1} +$$

$$\frac{c_k^2 e_k^2}{\sigma_k^2}\left(\delta_{k-1}^T \Lambda \delta_{k-1} + \gamma \lambda^2 \|\delta_{k-1}\|^2\right). \tag{4.41}$$

Next, we can obtain

$$e_k^2 + e_k \tilde{y}_{k-1}^T \delta_{k-1} = -\zeta_{k-1}^T \delta_{k-1} \leq \lambda \eta \|\delta_{k-1}\| \|e_k\|. \tag{4.42}$$

Substituting this and the definition of c_k, yields the final expression

$$\Delta V_k \leq -\frac{2c_k^2 e_k^2}{\sigma_k} < -\frac{c_k^2 e_k^2}{\sigma_k}. \tag{4.43}$$

It indicates that $\hat{y}_k$ and $\hat{\eta}_k$ are bounded. A detailed convergence analysis of the error e_k can be found in Verma $et\ al.$ [199].

4.4 RESULTS AND ANALYSES

In order to evaluate the effectiveness of the proposed method of controlling the depth of a gliding underwater robot, we conducted both simulations and aquatic experiments. The simulations were carried out in MATLAB®/ Simulink using the full-state dynamic model.

4.4.1 Depth Control in Gliding Motion

4.4.1.1 Simulation Results and Analysis

Extensive cases of depth control with various parameters were simulated, in which the target depth was set as 0.6 m. There were some parameters that needed to be set manually in our depth system, such as predictive steps N_p, control steps N_c, weight coefficients Q and R, reference tracking coefficients c_1 and c_2, and so on. Consequently, various simulations were conducted to compare the performances under different parameters, which also provided a valuable reference for the parameters of the following aquatic experiments. First, Q and R have a decisive influence on the results of depth control, in which a larger Q leads to a smaller steady error and a smaller R means a more violent control signal. However, we have employed a PID controller to make injector work on speed mode, which greatly reduced the failure rate of mechanical damage. Therefore, through the comparison of simulations, we set Q and R as 20 and 0.2, respectively. Second, N_c was set as the half of N_p in the depth system, and c_2 was set as 0.4 in simulations. Regarding the other model parameters, they were tabulated in Table 4.1. Furthermore, we just present the results under different N_p and c_1 in this section, as shown in Figure 4.5.

TABLE 4.1   Model Parameters of the Depth Control System.

Parameters	Value	Parameters	Value
m_{11}	11 kg	d_{11}	17 kg/s
m_{22}	13 kg	d_{22}	11 kg/s
m_{33}	0.6 kg	d_{33}	1.5 kgm²/s

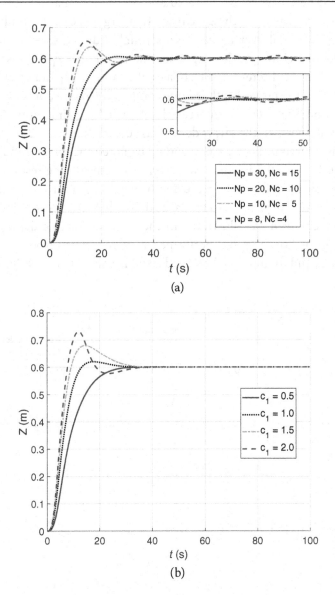

FIGURE 4.5   Simulations of depth control with parameters. (a) Results under different N_p. (b) Results under different c_1.

In Figure 4.5, results under four combinations of N_p and N_c are offered. Specially, smaller N_p may cause the overshoot owing to lack of ability to predict the future. However, large predictive steps will increase the computational burden. Through evaluating the results from the aspects of both overshoot and response time, it can be seen that the robot performed better when N_p were 20 and 30. In a similar way, by setting different c_1, we can draw the conclusion that larger c_1 can also lead to overshoot. In essence, larger c_1 means to set higher diving speed to be reference tracking. However, the response time with larger c_1 is not short due to the overshoot. Therefore, the choices of c_1 and c_2 depend on the target depth. When the target depth is large enough, the corresponding value of c_1 should be larger. Furthermore, a depth-switching simulation was also carried out, in which the target depth was successively set as 2 m, 4 m, and 2 m, as is figured in Figure 4.6. It should be noted that N_p, c_1, and c_2 were set at 30, 1.5, and 0.5 in this simulation, respectively. The result of depth shows no overshoot. Actually, the depth switching experiments are more suitable to large depth due to the large delay characteristics of buoyancy-driven mechanism. As illustrated in the starting stage of depth switching of Figure 4.6, the robot performed relatively slow response, and the pitch angle and diving speed reached a stable value after a while. Besides,

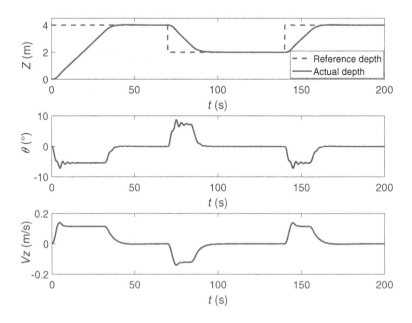

FIGURE 4.6 Results of the depth switching in simulation.

the maximum diving speed in Figure 4.6 is approximately 0.1 m/s, which offers the parameter v_{max} in heading control.

4.4.1.2 Experimental Results and Analysis without a Slider

We first obtained a relatively optimal parameter set based on a variety of simulations and applied these parameters to experiments based on minor adjustments. Therefore, the model and control parameters above were employed to test the depth control system in aquatic experiments. The difference lied in that the heading control worked. Moreover, N_p and N_c were set as 30 and 15 in this section, respectively.

First, the basic experiments of depth control without a movable slider were tested. Via setting the target depth as 0.6 m, the gliding underwater robot has successfully realized the depth control, the snapshot of which is figured in Figure 4.7. Meanwhile, the depth, pitch angle, and control signal are illustrated in Figure 4.8. Specially, we set c_1 as 0.2 due to the small target depth, which signified that the control signal in the experiments was small. The advantage of that is the result curve shows very smooth with no overshoot. As for the c_2 and $d_{threshold}$, they were set as 1.1 and 0.05, respectively. As was demonstrated, the robot arrived the target depth when

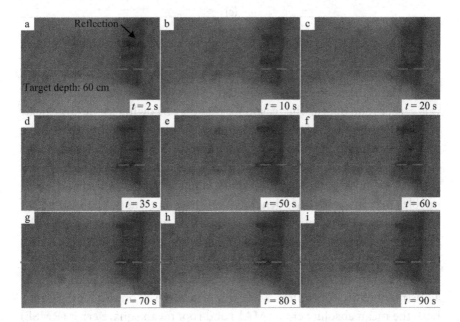

FIGURE 4.7 Snapshot sequences of depth control without a slider. (a) $t = 2$ s, (b) $t = 10$ s, (c) $t = 20$ s, (d) $t = 35$ s, (e) $t = 50$ s, (f) $t = 60$ s, (g) $t = 70$ s, (h) $t = 80$ s, (i) $t = 90$ s.

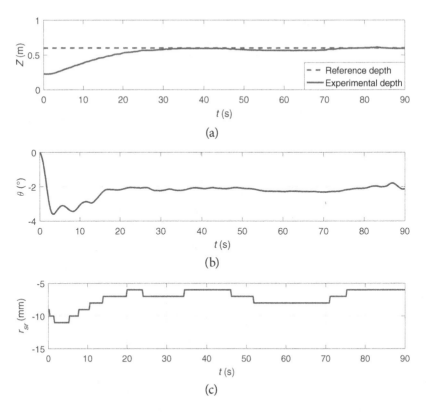

FIGURE 4.8 Experimental results of basic depth control. (a) Depth. (b) Pitch angle. (c) Control signal.

$t = 30$ s, and the pitch angle achieved the steady state when $t = 20$ s, which was earlier than depth. The reason for that may be owing to the large delay system. Besides, the robot had a slight upward trend in about 50 s. There may be two main reasons that contribute to the phenomenon. On one hand, there was a slight water flow disturbance in the underwater environment, which may make the robot shake slightly. On the other hand, some bubbles may be sometimes generated on the robot due to the insufficient smoothness of the robot shell, which could lead to the change of whole displacement. Certainly, the ability of the robot to resist these effects is also insufficient since the robot has a small overall weight and net mass. When the depth error increased, the robot went back to the target depth with the adjustment of the controllers. More important, based on the depth results, both the mean absolute error (MAE) and root mean square error (RMSE) were analyzed to discuss the control accuracy and performance of the proposed control method. When the gliding underwater robot entered the steady state at $t = 30$ s, the RMSE and the MAE of control error were 2.1 cm and 1.67 cm, respectively, which demonstrated the effectiveness of depth control. Note that we manually set a bias for the control signal to ensure that the robot remained hovering when there was no control, that was, the total displacement was equal to the total weight of the robot. This bias can be used to balance the small volume changes due to external disturbances, such as bubbles.

Moreover, the performances of heading control and SMO were also investigated, the results of which are shown in Figures 4.9 and 4.10. In experiments, we set the initial yaw angle as the target, which meant that the objective was to keep the heading of the robot unchanged. In essence, the basic principle of the regulation method is to use differential actions of the flippers to generate asymmetric hydrodynamic forces. For instance, when the robot deviates to the left direction, the left flipper will deflect while the right flipper keeps still. In aspect of heading control process of Figure 4.9, we can see the robot tended to depart from the target when the speed started to increase. Afterward, the flippers started to deflect, and the flippers' angles were calculated by u_f, which was related with speed. Consequently, the yaw angle gradually returned to the target yaw. More important, it should be noted that the heading control was only executed during a part of the control process. When the distance from the target depth was less than a certain threshold, heading control would be suspended. The reason for this design is that the robot almost has no speed in the final stage of depth control, so the heading control with deflecting method is unnecessary to exert. Regarding the SMO results plotted in Figure 4.10, the estimated diving speed is approximately consistent with the real one. Meanwhile, the forward speed is estimated to be near zero, which can be seen that the robot almost dived just along the vertical plane in snapshot sequences. As a matter of fact, the heading control and SMO are applied to assist in depth control. Therefore, the accuracy is within an acceptable range. Furthermore, by decreasing the model error between simplified model and actual plant model, the accuracy of SMO can be improved.

In order to further validate the effectiveness of the proposed method, we conducted more extended aquatic experiments. Figure 4.11 illustrates the depth control with better performance than that in basic experiments. In the first experiment, the RMSE and the MAE of depth error were 2.0 cm and 1.82 cm, respectively. Moreover, the RMSE and the MAE of depth error in the second experiment were 0.68 cm and 0.63 cm, respectively,

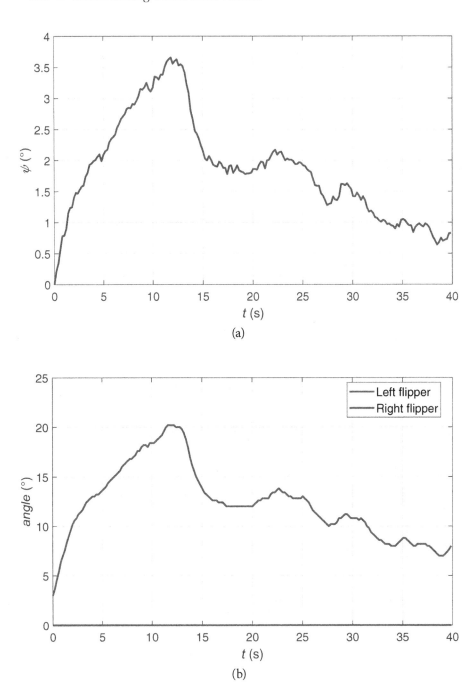

FIGURE 4.9   Heading control results of basic depth control. (a) Yaw angle. (b) Flippers' angle.

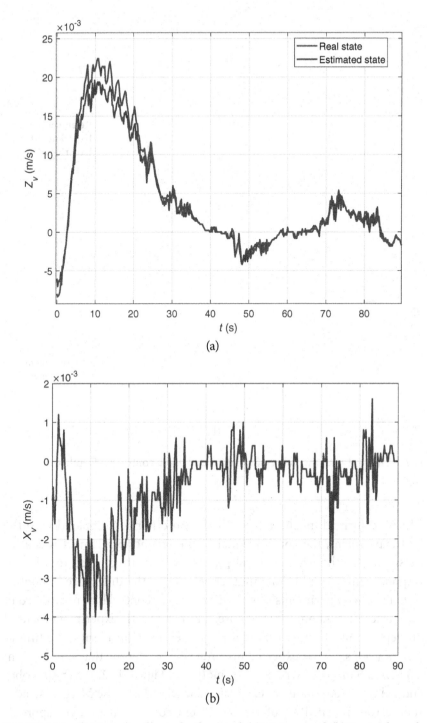

FIGURE 4.10 SMO results of basic depth control. (a) Diving speed. (b) Forward speed.

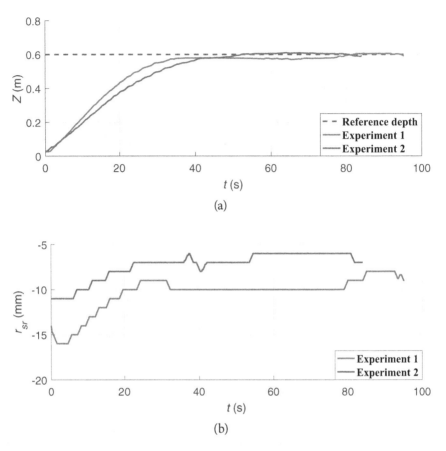

FIGURE 4.11 Experimental results of extended depth control. (a) Depth. (b) Control signal.

which performed much better in both two criteria. Compared with the first experiment, the control signal of the second experiment was a little smaller in the whole control process, which directly led to a longer response time. In addition, another reason for the difference in control signal in two experiments was the difference in control bias mentioned in the earlier discussion. Furthermore, another aquatic experiment focused on depth control under different target depths. Considering the limitation of pool size, the experiments under target depths that ranged from 0.3 m to 0.5 m were carried out, as plotted in Figure 4.12. When the robot entered the steady state, we calculated the RMSE and the MAE of steady-state error. The RMSEs of the 30–50-cm control target were approximately 3.56 cm, 3.3 cm, and 3.35 cm, respectively, while their MAEs were

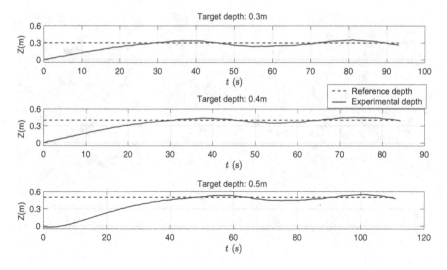

FIGURE 4.12   Depth results under different target depth.

approximately 3.13 cm, 2.86 cm, and 2.87 cm, respectively. More careful inspection shows that the error has slight undulation. There may be
two reasons for this phenomenon, for example, small target depth and
unsuitable parameters of the controller. Since the delay characteristics,
the response of the robot is not sensitive. Especially when the target depth
is small, the robot is easier to perform overshooting. As a result, it is more
suitable for the gliding underwater robot to realize depth control under
large target depth.

4.4.1.3 Experimental Results and Analysis with a Slider

Moreover, we also investigated the effect of a movable slider during the
depth-control process. In experiments, we set the position of the slider
from 10 mm to 30 mm, and the heading control was also employed. In
particular, it should be noted that we moved the slider to the target position at the initial stage and moved it back to zero when the robot arrived
the 0.3-m depth, which contributed to the steady depth control in the final
stage. In Figure 4.13, the snapshot sequences of depth control experiments
under three slider positions are illustrated, and the lateral and longitudinal distances of the gliding are marked out, which shows the orders of
the lateral distance is $x_{20mm} > x_{10mm} > x_{30mm}$. Figures 4.14 and 4.15 reveal
the depth and angle under different slider positions, respectively. Besides, the
results of depth and yaw angle validate that the robot can successfully

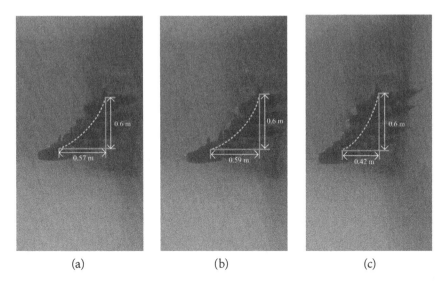

(a)   (b)   (c)

FIGURE 4.13   Snapshot sequences of experiments under different slider positions. (a) r_m = 10 mm. (b) r_m = 20 mm. (c) r_m = 30 mm.

realize the depth and heading control with a slider. In particular, in the initial stage of Figure 4.14, the diving speed with r_m = 30 mm was the smallest, the reason for which may be that the attitude of the body was not adjusted to the stable state in time. This point can also be verified by pitch angle in Figure 4.15. Therefore, the robot with r_m = 30 mm was late to reach a depth of 0.3 m. In addition, it can be found that the robot with r_m = 20 mm was the first one to achieve the 0.3-m depth at t = 17 s. Hence, its pitch angle came back much earlier, which contributed to its large horizontal distance due to a small gliding angle. Moreover, the moving of the injector piston may also make some influence on the attitude and depth results since it can also change the center of gravity. For instance, the robot with r_m = 30 mm remained for a relatively long time in the state of large r_{sr}, which signified that the injector piston had more effect on the gravity's center shift. Therefore, we can draw the conclusion to some extent that the robot can achieve different gliding angles by moving the slider during the depth control process, which can be used to further control the horizontal gliding distance.

4.4.1.4 Discussion

Due to the large delay and poor maneuverability of buoyancy-driven underwater robot, it is hard to achieve motion control with high precision.

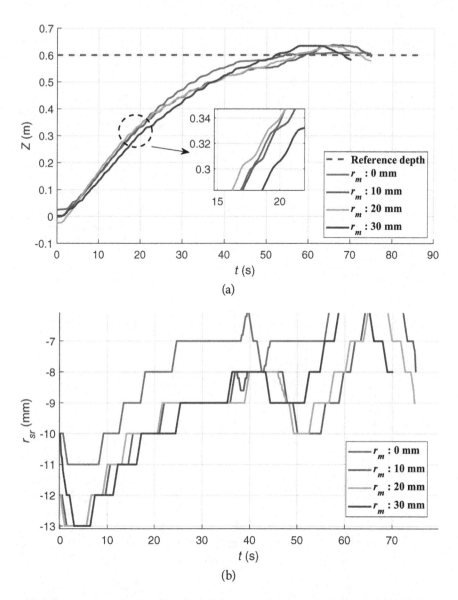

FIGURE 4.14 Experimental results of depth control under different r_m. (a) Depth. (b) Control signal.

The gliding underwater robot, as a hybrid underwater robot, is particularly designed to overcome the weaknesses through combining the robotic dolphin and underwater gliders. Therefore, in order to explore the performance of the gliding underwater robot in vertical plane, high precision depth control only by an injector-based buoyancy mechanism has been successfully

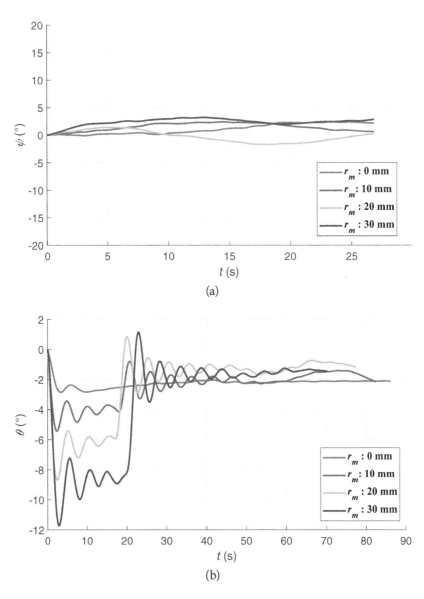

FIGURE 4.15 Angle results of depth control under different r_m. (a) Yaw angle. (b) Pitch angle.

realized in this chapter. Meanwhile, the effects of movable sliders for gliding angle in the depth control process are also investigated, which not only validates the feasibility of the platform and proposed control framework but also accumulates valuable engineering experience for ocean exploration. Compared with depth control of traditional underwater robots [187], the gliding underwater robot has more excellent maneuverability due to the rotatable surfaces, for example, the heading control is achieved via deflecting the flippers. Compared with the results reported by Makrodimitris *et al.* [136], although the size of our robot is much larger than theirs, we still achieve a better performance for the aspect of depth error. Additionally, an MPC controller, a heading controller, and an SMO are programmed by C language and applied to an embedded platform with a float-point unit (FPU). In order to reduce the computing burden, the dynamic model is simplified. Hence, the real-time performance is improved to some extent.

In spite of successful depth control exerted on the gliding underwater robot, there are some limitations of some aspects. First, since the injector changes buoyancy by sucking and draining, it needs to overcome water pressure when draining, which means it is much more difficult to accomplish the buoyancy adjustment in a deepwater environment. To address this issue, a high-power motor or internal liquid bladders can be applied, yet the latter design needs enough internal space. Second, due to the simplified model error and inaccurate parameters, the estimation accuracy of SMO is not particularly high. Through enhancing the model accuracy or using another estimation method, such as the Kalman filter, the problem can be improved. Third, regarding real-time computing, the calculated control signal seems a little unsmooth. By carrying a more efficient master chip or appropriately decreasing predictive steps, the MPC can run faster. The improvements for state estimation and faster computing are ongoing endeavors.

4.4.2 Depth Control in the Dolphin-Like Motion

4.4.2.1 Control Parameter Optimization

For the ACA-based controller, we needed to find optimal values for four control parameters, namely, q_{min}, λ, γ, and μ. Hence, we needed to design a suitable objective function to solve this problem. The most important factor for depth control is the error (accuracy), followed by the transition time. Thus, we created an objective function with parameter constraints based on the error $|e|$ and transition time t_s, as follows:

$$arg\min \quad \kappa_1|e| + \kappa_2 t_s$$
$$s.t. \begin{cases} \kappa_1 + \kappa_2 = 1 \\ 0 < \kappa_1, \kappa_2 < 1 \\ 0 < q_{min}, \mu < 5 \\ 0 < \lambda, \gamma < 10 \end{cases} \qquad (4.44)$$

Here, since the dimensions of the two optimization goals were different, we used min–max normalization to normalize the units by setting bounds on their values. In addition, by adjusting the κ_1 and κ_2 values, we could focus the optimization process more on the error or transition time. Since the dynamic model was nonlinear, we used MATLAB's genetic algorithm toolbox to search for an optimal set of control parameters.

By setting $\kappa_1 = 0.9$ and $\kappa_2 = 0.1$, we obtained the following optimal control parameter set: $q_{min} = 1.5$, $\lambda = 3.51$, $\gamma = 4.19$, and $\mu = 1.59$. Then, we made some minor adjustments to this parameter set based on the results of the aquatic experiments. The final parameters used were $q_{min} = 0.7$, $\lambda = 3.51$, $\gamma = 4.19$, and $\mu = 0.99$ for controlling ψ, and likewise for controlling β except that $q_{min} = 1.2$. Here, we found that the control parameters calculated using our full-state dynamic model provided accurate initial values for actual operation, confirming the model's accuracy.

4.4.2.2 Results for Flipper-Based Control

First, we used the previously mentioned control parameters to test controlling the dolphin's depth by adjusting the flipper deflection angle ψ. Here, we bounded the control signal u_k to lie within the range $[-45°, 45°]$ based on mechanical constraints. Then, to generate dolphin-like motion, we set the CPG frequency to 1 Hz, the intrinsic amplitudes of the waist and caudal joints to $20°$ and $30°$, respectively, and the phase difference between the two joints to $45°$.

In this experiment, we set the target depth to 30 cm. Figure 4.16 shows a sequence of snapshots taken during the control process, which lasted for approximately 24 s. Figure 4.17 shows the real-time depth data collected by the onboard sensor. Here, we see that the experimental and simulated results are approximately consistent. In order to better analyze the control accuracy and controller performance, we calculated both the MAE and the RMSE based on the depth results. Once the robotic dolphin had entered a steady state, the MAE and the RMSE of the control error d_e were 0.43 and 0.09 cm, respectively, demonstrating that the controller was effective. However, the transition time was somewhat long, mainly because the dolphin's forward speed was not high enough for the flippers to rapidly generate the desired pitch moments.

Figure 4.18 plots the flipper and pitch angles during the experiment, based on the recorded IMU data and control signal. The pitch angle has been filtered, since the dolphin's head generated significant pitch jitter (approximately $\pm 8°$) due to flapping of the waist and caudal joints, and these oscillations seriously affected the control process. To combat this,

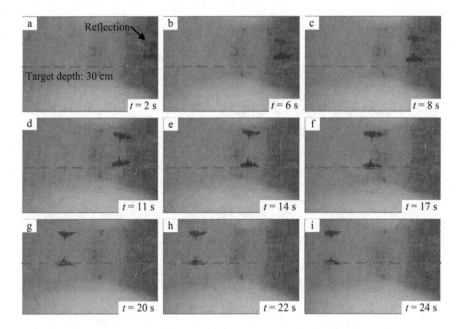

FIGURE 4.16 Snapshot sequence for flipper-based depth control. (a) $t = 2$ s, (b) $t = 6$ s, (c) $t = 8$ s, (d) $t = 11$ s, (e) $t = 14$ s, (f) $t = 17$ s, (g) $t = 20$ s, (h) $t = 22$ s, (i) $t = 24$ s.

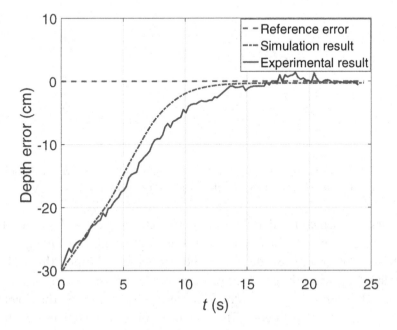

FIGURE 4.17 Depth error for flipper-based depth control.

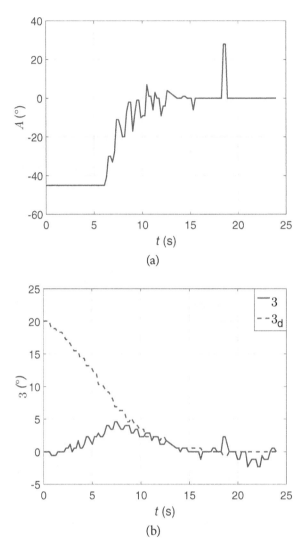

FIGURE 4.18   Control inputs and outputs during the aquatic experiments, show-ing the (a) flipper angle and (b) target and actual pitch angles.

we first performed median filtering on the pitch angle based on the CPG frequency, then used the filtered data θ for depth control. Although such filtering can introduce a delay into the attitude signal input, this had no obvious impact on the depth control process.

In addition, as Figure 4.18b illustrates, θ_d decreased as the dolphin approached the target depth. The control input ψ was fixed at $-45°$ until $t = 6$ s and then began to increase as the pitch angle θ approached the target θ_d. However, we should also note that θ only gradually began to track θ_d after $t = 6$ s, and the maximum pitch angle achieved was only around 5°, which also explains the insufficient pitch moment. In addition, we can see a fluctuation at about 18 s, when the actual depth became slightly offset from the target depth during the depth maintenance phase, as can be seen in Figure 4.17.

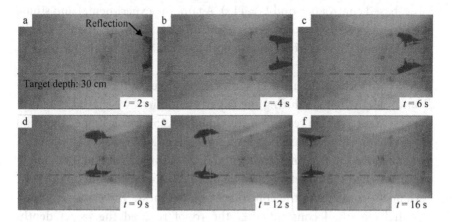

FIGURE 4.19   Snapshot sequence for CPG-based depth control. (a) $t = 2$ s, (b) $t = 4$ s, (c) $t = 6$ s, (d) $t = 9$ s, (e) $t = 12$ s, (f) $t = 16$ s.

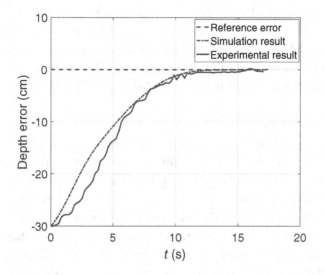

FIGURE 4.20   Depth error for CPG-based depth control.

4.4.2.3 Results for CPG-Offset-Based Control

The second experiment focused on controlling the depth using the CPG offset β_i. Here, we bounded β_i to lie within the range $[-18°, 18°]$. Figures 4.19 and 4.20 show a snapshot sequence and depth data recorded during the control process, respectively. Compared with ψ-based control, this approach had two advantages. First, the robot was more stable, with no fluctuations after it entered the depth maintenance phase. Second, the transition time (approximately $t_s = 10.3$ s) was significantly shorter than for ψ-based control (around $t_s = 14$ s). Again, the experimental and simulated results were in good agreement. In addition, the MAE and the RMSE were 0.40 and 0.08 cm, respectively, both lower than the results for flipper-based control.

More important, Figure 4.21 shows the changes in the control variable β_i and the pitch angle during the experiment. First, we should note that the actual pitch angle reached a maximum of $25°$, much higher than the angle seen for flipper-based control, indicating that β_i-based control can provide more suitable pitch moments. In addition, the actual pitch angle curve matched the target pitch angle more closely during the diving process, again confirming the controller's effectiveness. The error remained essentially a small constant after the robot reached the target depth, because we included a control threshold in the control process. Specifically, control was not applied when the depth error was below a threshold of $e_s = 0.8$ cm. It is also worth noting that the pitch angle error indicates the robot did not dive in exactly the direction indicated by the pitch angle, due to mechanical inaccuracies. In addition, compared with the results reported by Shen et al. [190], we achieved equal depth accuracy and a similarly short transition time by controlling β_i instead of ψ. However, we were able to use a lower frequency (1 Hz vs. 1.5 Hz), which could save energy.

4.4.2.4 Discussion

We have implemented two methods of controlling the depth of a gliding underwater robot, not only confirming the feasibility of the platform and control algorithm but also gathering valuable engineering experience for ocean exploration. Compared with the method used by Verma et al. [199], we were able to calculate the control parameters more accurately using our full-state dynamic model and achieved a smaller steady-state error. In addition, our approach involves fewer control parameters than the method used by Yu et al. [190], which needs to consider not only the SMC model's parameters but also fuzzy rules. Our simulations and aquatic experiments

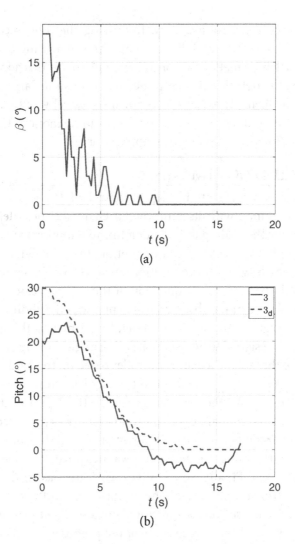

FIGURE 4.21 Control inputs and outputs during the aquatic experiments, show-ing the (a) flipper angle and (b) target and actual pitch angles.

have shown that we can see that controlling β_i can achieve a similar, even better performance than controlling ψ. In particular, controlling β_i can free the flippers from having to control the pitch moment, leaving them to focus on the more important task of adjusting the yaw moment, as the only structure that can handle this. This potentially suggests new ideas for 3-D path tracking.

Nevertheless, our system does have some limitations. First, the robotic dolphin did not swim exactly along the direction given by the pitch angle, which may affect control accuracy. Improving the robot's construction could reduce this effect, and we could also consider using depth data as direct feedback information for control. Second, robots that adopt dolphin-like motions to control their depth inevitably consume large amounts of energy due to their propulsion mechanism. A novel way to address this issue would be to combine gliding and dolphin-like motions to extend the robot's operational time in practical applications.

4.5 CONCLUDING REMARKS

In this chapter, aiming at the depth control of the gliding underwater robot, combined with its multimodal characteristics, depth control frameworks for the gliding and the dolphin-like motions were proposed, respectively. For the aspect of gliding motion, first, the dynamic model is simplified, which saves the computing resources of the embedded platform. Second, the real-time acquisition of the gliding velocity is realized by using the sliding mode observer, which provides the gliding state quantity for the design of the depth controller. By estimating the speed, a yaw controller was designed to adjust the heading angle. Finally, in view of the large delay of the system, a depth controller based on the model prediction algorithm was proposed, and a speed planner based on the Bezier curve was presented to solve the overshoot issue. In the aspect of the dolphin-like motion, a dual-modal depth control method was proposed. First, the LOS navigation method was used to convert the depth control into pitch control. Second, the dual-mode selector was designed, followed by its basic principle analyzed. Finally, an adaptive algorithm was used to construct the pitch control law, and the offline optimization of the control parameters was realized based on the integrated dynamic model. Simulation and experiments verified the effectiveness of the proposed methods for depth control in the gliding and dolphin-like motions.

REFERENCES

[185] J.-H. Li and P.-M. Lee, "Design of an adaptive nonlinear controller for depth control of an autonomous underwater vehicle," *Ocean Eng.*, vol. 32, nos. 17–18, pp. 2165–2181, 2005.

[186] C. Silvestre and A. Pascoal, "Depth control of the INFANTE AUV using gain-scheduled reduced order output feedback," *Control Eng. Pract.*, vol. 15, no. 7, pp. 883–895, 2007.

[187] H. Wu, S. Song, K. You, and C. Wu, "Depth control of model-free AUVs via reinforcement learning," *IEEE Trans. Syst. Man Cybern. Syst.*, vol. 49, no. 12, pp. 2499–2510, Dec. 2019.

[188] N. A. A. Hussain, M. R. Arshad, and R. Mohd-Mokhtar, "Underwater glider modelling and analysis for net buoyancy, depth and pitch angle control," *Ocean Eng.*, vol. 38, no. 16, pp. 1782–1791, 2011.

[189] J. Yu, F. Sun, D. Xu, and M. Tan, "Embedded vision-guided 3-D tracking control for robotic fish," *IEEE Trans. Ind. Electron.*, vol. 63, no. 1, pp. 355–363, Jan. 2016.

[190] J. Yu, J. Liu, Z. Wu, and H. Fang, "Depth control of a bioinspired robotic dolphin based on sliding-mode fuzzy control method," *IEEE Trans. Ind. Electron.*, vol. 65, no. 3, pp. 2429–2438, Mar. 2018.

[191] F. Shen, Z. Cao, C. Zhou, D. Xu, and N. Gu, "Depth control for robotic dolphin based on fuzzy PID control," *Int. J. Offshore Polar Eng.*, vol. 23, no. 3, pp. 166–171, 2013.

[192] Z. Li, J. Deng, R. Lu, Y. Xu, J. Bai, and C.-Y. Su, "Trajectory-tracking control of mobile robot systems incorporating neural-dynamic optimized model predictive approach," *IEEE Trans. Syst. Man Cybern. Syst.*, vol. 46, no. 6, pp. 740–749, Jun. 2016.

[193] F. Ke, Z. Li, H. Xiao, and X. Zhang, "Visual servoing of constrained mobile robots based on model predictive control," *IEEE Trans. Syst. Man Cybern. Syst.*, vol. 47, no. 7, pp. 1428–1438, Jul. 2017.

[194] M. Yue, C. An, and Z. Li, "Constrained adaptive robust trajectory tracking for WIP vehicles using model predictive control and extended state observer," *IEEE Trans. Syst. Man Cybern. Syst.*, vol. 48, no. 5, pp. 733–742, May 2018.

[195] C. Shen, Y. Shi, and B. Buckham, "Path-following control of an AUV using multi-objective model predictive control," in *Proc. Amer. Control Conf.*, Boston, MA, USA, Jul. 2016, pp. 4507–4512.

[196] C. Shen, Y. Shi, and B. Buckham, "Nonlinear model predictive control for trajectory tracking of an AUV: A distributed implementation," in *Proc. IEEE Conf. Decis. Control*, Las Vegas, NV, USA, Dec. 2016, pp. 5998–6003.

[197] D. C. Fernández and G. A. Hollinger, "Model predictive control for underwater robots in ocean waves," *IEEE Robot. Autom. Lett.*, vol. 2, no. 1, pp. 88–95, Jan. 2017.

[198] A. J. Ijspeert, A. Crespi, D. Ryczko, et al., "From swimming to walking with a salamander robot driven by a spinal cord model," *Science*, vol. 315, pp. 1416–1420, 2007

[199] S. Verma, D. Shen, and J. Xu, "Motion control of robotic fish under dynamic environmental conditions using adaptive control approach," *IEEE J. Ocean Eng.*, vol. 43, pp. 381–390, 2018.

CHAPTER 5

Heading and Pitch Regulation of Gliding Motion Based on Controllable Surfaces

5.1 INTRODUCTION

In fact, gliding motion has been realized by underwater gliders for long endurance a long time ago. A variety of underwater gliders have been developed and successfully applied in ocean observatories and exploration [200]. Relying on this energy-saving gliding style, these underwater gliders can easily travel up to thousands of kilometers or for several months in oceans. However, most underwater gliders own obvious drawbacks, for example, much lower speed and lower maneuverability, due to this specific buoyancy-driven mode. Fortunately, the gliding robotic dolphin addresses these deficiencies, since it can implement high speed and high maneuverability by imitating a dolphin's dorsoventral propulsion. Not only that, but as a hybrid-driven one, the gliding robotic dolphin can also easily and effectively integrate these two propulsive modes for better locomotion. For instance, the movable flippers and fluke are not only utilized to generate propulsive forces in dolphin-like swimming mode but also acted as control surfaces for attitude regulation in the gliding motion. Regarding the traditional underwater gliders, they have fixed wings for gliding but only

DOI: 10.1201/9781003347439-5

build internal movable masses for attitude regulation, which are made up of rechargeable batteries in most cases [44].

Heading regulation is crucial for navigation during gliding. So is heading stabilization, since factors like environmental disturbances and asymmetry of hydrodynamic structures would cause gliders to depart from the desired direction. In general, existing approaches for adjusting heading angles can be categorized into two classes by actuators: rolling by internal movable masses to cause a banked turn [15], turning by vertical rudders [17]. Concretely, as for the former, the pitch moment, lift, and drag forces are rotated out of the vertical plane, and thereby their lateral components arise and compel gliders to turn. Regarding the latter, yaw movements come from the hydrodynamic moment caused by deflection of rudders. The former costs no effort on dynamic sealing of driving shaft and reduces the risk of leakage, while the direction of roll should be changed during infections between upward and downward glide, which would cause frequent actions of the mass actuators in shallow-water operations involving many inflections [44]. By comparison, the latter conserves the interior space of gliders, and the relation between rudder deflection and heading direction is fixed. From the viewpoint of control methods, proportional-integral-derivative (PID) feedback controllers are commonly used for their simplicity and convenience [15]. However, little effort has been devoted to the research of heading control via differential actions of horizontal control surfaces up to now, since traditional gliders have nearly no movable horizontal surfaces for attitude control. Therefore, heading control of new-style gliders with movable horizontal surfaces, such as gliding robotic dolphins and flying wing gliders [201], is a new problem.

Pitch is generally regulated via shifting the internal moveable mass fore and aft within the glider, while heading control is achieved through rotating an eccentric mass about the glider's longitudinal axis to drive the glider to roll. To achieve an accurate performance, many closed-loop control approaches are proposed [202]–[203]. Cao *et al.* proposed a nonlinear multiple-in, multiple-out (MIMO) adaptive backstepping control scheme to control an underwater glider in sawtooth motion [124]. In these control methods, the objectives are all the internal movable mass. This is different from the gliding robotic dolphin, which owns several control surfaces including flippers and fluke for attitude regulation. As an example, Zhang *et al.* presented the passivity-based stabilization of underwater gliders with a whale tail–like control surface [129]. However, the effect of the horizontal fluke on the gliding motion has not been discussed. Owing to the large force arm of the fluke, the produced pitch moment should be very large, which will cause a series of changes, such as pitch angle, gliding angle, and gliding velocity. Therefore, it is worth exploring how the fluke affects the gliding motion.

Further progress is made in this chapter, whose objective is to achieve the attitude control for an underwater gliding robot in the gliding motion. The contributions made in this chapter are summarized as follows:

- By virtue of experimental data, the regulating effect of the controllable fins on gliding motion is investigated, including the yaw and pitch motions.

- An attitude control framework is proposed to achieve the heading and pitching control, including the backstepping method, sliding-mode observer (SMO), and solver.

- Extensive simulations and aquatic experiments are carried out. The obtained results validate the effectiveness of the proposed methods.

The remainder of this chapter is organized as follows. Section 5.2 offers a gliding analysis under movable fin. The control methods of yaw and pitch movement are elaborated in Section 5.3. Simulation and experimental analysis are provided in Sections 5.4 and 5.5, respectively. Finally, Section 5.6 provides our conclusions.

5.2 GLIDING ANALYSIS UNDER MOVABLE FIN

The 1.5-m gliding robotic dolphin has been selected to verify our method, as shown in Figure 5.1.

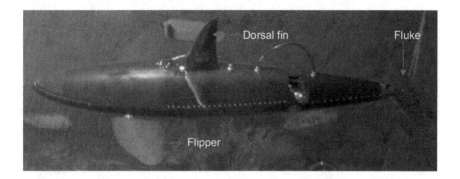

FIGURE 5.1 Overview of the developed gliding robotic dolphin.

5.2.1 Analysis of Yaw Movement

The yaw moment can be generated through the differential movement of the pectoral fins, allowing three-dimensional (3-D) spiral motion. Keep the bias angle of the right pectoral fin as $0°$ and change the bias angle of the left pectoral fin, then the state data of the spiral dive can be obtained, as shown in Figures 5.2 to 5.5 and Table 5.1.

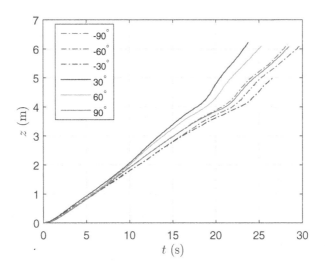

FIGURE 5.2   Depth values under different bias angles of the left pectoral fin.

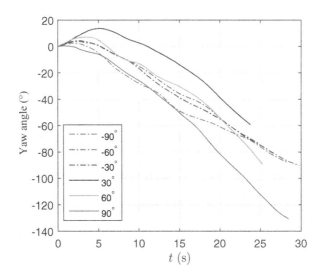

FIGURE 5.3   Yaw angles under different bias angles of the left pectoral fin.

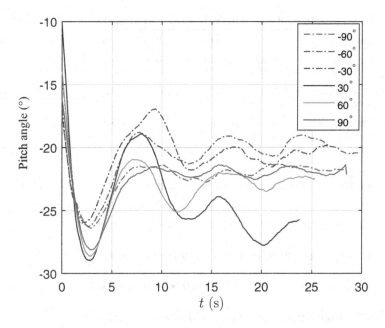

FIGURE 5.4   Pitch angles under different bias angles of the left pectoral fin.

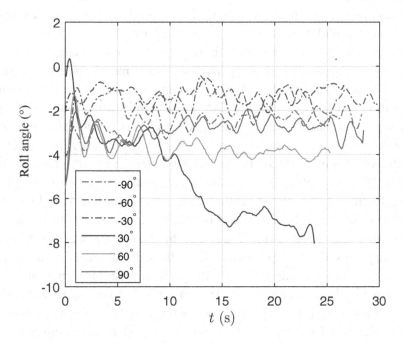

FIGURE 5.5   Roll angles under different bias angles of the left pectoral fin.

TABLE 5.1   Experimental Results under Different Bias Angles of the Left Pectoral Fin

Bias angles (°)	−90	−60	−30	30	60	90
Average diving velocity (m/s)	0.217	0.206	0.188	0.262	0.241	0.214
Average yaw angular velocity (°/s)	−3.1	−3.0	−3.0	−2.5	−3.5	−4.6
Stable yaw angular velocity (°/s)	−3.6	−4.1	−4.0	−5.5	−6.6	−6.2

From the stable-state data, the following phenomena can be found:

1. Yaw angular velocity: $-90° \approx -60° \approx -30° < 30° < 90° < 60°$

2. Diving velocity: $-30° < -60° < -90° \approx 90° < 60° < 30°$

3. Pitch angle: $30° < 60° < 90° \approx -90° < -30° < -60°$

4. Roll angle: $30° < 60° < 90° \approx -90° < -60° < -30°$

The yaw angular velocity depends on the yaw moment generated by the left pectoral fin. Meanwhile, the yaw moment depends on the gliding velocity and the angle of attack (AoA) of the left pectoral fin. When the bias angle $\delta_l = 90°$, the gliding velocity is compromised and the AoA (or the water-facing surface) of the left pectoral fin is larger. As a result, a larger left-hand resistance is generated, thus the yaw moment is more prominent. When $\delta_l = 30°$, due to the small AoA, the yaw velocity is small at the beginning. But this is precisely because the AoA is small. The resistance to the gliding movement is relatively small, so the gliding velocity is faster after entering the stable state, which will also produce larger yaw moment. Therefore, the stable yaw angular velocity is significantly faster than that in the starting stage. When $\delta_l = 60°$, the AoA is larger than that of $\delta_l = 30°$ and the gliding velocity is faster. Therefore, the yaw moment is the largest and the yaw angular velocity is the fastest in this case. When $\delta_l < 0$, the AoA is larger than that of $\delta_l > 0$, and the drag caused by the pectoral fins is too large. Consequently, the gliding velocity is little; thus, the difference of yaw angular velocity is not obvious. Besides, the pitch angle and roll angle at $\delta_l > 0$ are smaller than those at $\delta_l < 0$. The reason is that the resultant force of the lift and drag generated by the pectoral fins at $\delta_l < 0$ has a component along the negative direction of the z-axis of C_b. Thus, a head-up moment will be generated, which will increase the pitch angle. Furthermore, this component will compel the robot to roll forward, increasing the roll angle. For the same reason, the dive speed at $\delta_l < 0$ is slower than that at $\delta_l > 0$.

It should be noted that the collected experimental data is discrete. The earlier analysis only compares the gliding state relationships under six groups of values, but the analysis method is applicable to all the values. The motion of the gliding robotic dolphin has significant underactuated characteristics, and the motion states are coupled with each other. The generation of the above gliding data is the result of the interaction of various factors. The yaw moment is closely related to the bias angle of the pectoral fins, the AoA of the robot, and the magnitude of the gliding velocity. The bias angle not only changes the yaw moment, but also changes other motion states, such as the pitch angle, roll angle, AoA, and gliding velocity. In turn, these states will affect the yaw moment until an equilibrium state is reached. Therefore, the value of δ_l at the extreme value of the yaw moment cannot be determined according to the comparison of the stable yaw angular velocity in the previously mentioned motion data. For example, although the stable yaw angular velocity at $\delta_l = -60^\circ$ is smaller than that at $\delta_l = 30^\circ$, the magnitude of the generated yaw moment of the former is greater when the gliding velocity and AoA are identical (e.g., $\alpha_b = 15^\circ$). Meanwhile, the former will cause the gliding velocity to be too low to acquire a high yaw angular velocity. When the gliding velocity and the AoA are the same, it is convenient to compare the yaw moment produced by different bias angles. Figure 5.6 shows the relationship among the yaw moment, the bias angle of the left pectoral fin, and the AoA at the same gliding velocity, which was generated using computational fluid dynamics (CFD) simulation data. It can be seen that there is a nonlinear relationship between them. In addition, the left pectoral fin is not only capable of producing leftward yaw moment but also can produce rightward yaw moment when the bias angle is near 0°. Since the δ_l selected in the experiment is relatively sparse, these phenomena have not been observed.

The following conclusions can be drawn from the earlier analysis:

1. In the 3-D gliding motion that the pectoral fins participate in, there exists a coupling phenomenon between states. When the pectoral fin changes the yaw angle, it also changes other states, such as gliding velocity, pitch angle, and roll angle.

2. The yaw moment has a nonlinear relationship with the bias angle and the AoA.

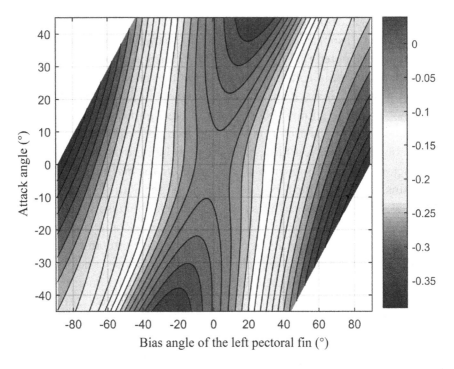

FIGURE 5.6 Relationship between the yaw moment, the bias angle of the left pectoral fin, and the AoA ($\|V_b\| = 0.4$ m/s and $\Omega_b = \mathbf{0}_{3 \times 1}$).

3. Under the same gliding velocity and AoA, the yaw moment and the average yaw angular velocity are not positively correlated. This is because the gliding velocity is also affected, changing the yaw moment.

Conclusion (3) brings the enlightenment to heading control: If small-angle yawing is performed, the bias angle of the pectoral fin with a large yaw moment can be used. If large-angle yawing is performed, it is more suitable to apply a bias angle with a large average yaw angular velocity.

5.2.2 Analysis of Pitch Movement

In this section, we conduct gliding equilibria analyses in different deflection angles of the controllable fins.

To describe the gliding robotic dolphin moving in 3-D space easily, we define several coordinate frames including an inertia frame (C_g), a body-fixed frame (C_b), and flipper and fluke-fixed frames, as shown in Figure 5.7.

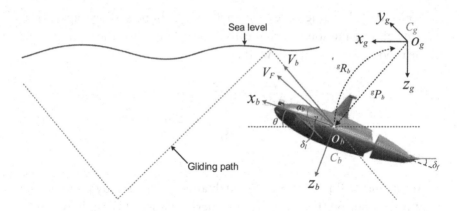

FIGURE 5.7   Definition of coordinate frames and notations.

Distinctly, the robot has no angular speed and acceleration in the steady gliding state. Meanwhile, both movable mass and oil bladders keep relatively static in the robot. According to the dynamic model, we can derive the planar gliding equilibria via deducing the equilibrium forces in x-and z-axis and the equilibrium moments around y-axis. Furthermore, we can easily derive the relationship between pitch angle θ and the α_j for the robotic dolphin as follows:

$$\theta = \arctan \frac{\sum_j \left[C_{j,l}'(\alpha_j)\sin\alpha_b - C_{j,d}'(\alpha_j)\cos\alpha_b \right]}{\sum_j \left[C_{j,l}'(\alpha_j)\cos\alpha_b + C_{j,d}'(\alpha_j)\sin\alpha_b \right]}, \tag{5.1}$$

where $j \in \{b, l, r, f\}$, which separately corresponds to the body, the left flipper, the right flipper, and the fluke. $C_{j,l}'(\alpha_j) = \frac{1}{2}\rho S_j C_{j,l}(\alpha_j)$ and $C_{j,d}'(\alpha_j) = \frac{1}{2}\rho S_j C_{j,d}(\alpha_j)$, where ρ is the water density. S_j means the reference area of the j part. $C_{j,l}(\alpha_j)$ and $C_{j,d}(\alpha_j)$ denote the lift coefficient and drag coefficients of the j part, which are affected by the AoA α_j, respectively.

As for the controllable fins, that is, the flippers and the fluke, the AoA α_j depends on both the AoA of body α_b and the deflection angle δ_j of the controllable fins.

$$\alpha_j = \alpha_b - \delta_j \quad (j \in \{l, r, f\}). \tag{5.2}$$

Notice that the same deflection angle of the flippers always appears in a steady gliding motion, which means $\delta_l = \delta_r$. Regarding the gliding angle of the robot $\gamma = \theta - \alpha_b$, it follows that

$$\gamma = -\arctan \frac{\sum_j C_{j,d}{'}(\alpha_j)}{\sum_j C_{j,l}{'}(\alpha_j)} \quad (j \in \{b, l, r, f\}). \tag{5.3}$$

According to Equation 5.1, the pitch angle θ in a steady gliding state mainly depends on the hydrodynamic performance of the dolphin body, the flippers and the flukes, such as the lift coefficient and drag coefficient of each part. In other words, the pitch angle θ is decided by the AoA α_j, since the robotic dolphin has a certain geometric shape. Therefore, we can easily regulate the pitch angle θ via the deflection angles of controllable flippers as well as the fluke, δ_j. Recalling Equation 5.3, we can get that the gliding angle γ is also decided by the hydrodynamic coefficients and affected by α_j of each part. Evidently, it has a strong relationship with the whole lift-to-drag ratio of the robotic dolphin. In detail, a larger lift-to-drag ratio always results in a smaller absolute value of the gliding angle, along with a longer distance in a whole circle of the gliding motion, that is, an upward-and-downward glide. This means that a large lift-to-drag ratio can be beneficial to improving the gliding efficiency (e.g., the gliding distance over one cycle of the gliding motion) of the robot. Therefore, the gliding robotic dolphin can easily obtain a better gliding efficiency through adjusting the deflection angles of the controllable flippers as well as the fluke.

Furthermore, the gliding velocity in the steady gliding motion can be derived as follows:

$$\left\| {}^g V_b \right\| = \sqrt{\frac{-m_0 g \sin \gamma}{\sum_j C_{j,d}{'}(\alpha_j)}} \quad (j \in \{b, l, r, f\}), \tag{5.4}$$

where $m_0 = \dfrac{\rho}{\rho_o}(\dfrac{1}{2}m_o - m_{ex})$ is the net buoyancy from the buoyancy-driven system. ρ_o is the oil density. m_{ex} is the oil mass in the external bladder.

Based on the preceding equations, we have

$$\left\| {}^{g}V_{b} \right\| = \left\{ |m_{0}| g \left\{ \left[\sum_{j} C_{j,d}'(\alpha_{j}) \right]^{2} + \left[\sum_{j} C_{j,l}'(\alpha_{j}) \right]^{2} \right\}^{-\frac{1}{2}} \right\}^{\frac{1}{2}}. \tag{5.5}$$

The velocities in the horizontal plane and vertical plane can be computed as follows:

$$\begin{cases} {}^{g}V_{bx} = \left\| {}^{g}V_{b} \right\| \cos\gamma \\ {}^{g}V_{bz} = - \left\| {}^{g}V_{b} \right\| \sin\gamma \end{cases}. \tag{5.6}$$

According to Equations 5.5 and 5.6, the gliding velocity of the robot can be analyzed in detail. First, we can find that the gliding velocity is decided by the net mass m_{0} and AoA α_{j}. When in a gliding equilibrium, a larger m_{0} contributes to a higher gliding velocity. Given a certain m_{0}, both horizontal and vertical velocities are decided by α_{j}, since the gliding angle is also a function of α_{j}. Furthermore, the AoA α_{j} is decided by an equilibrium moment, which takes the following form:

$$\tau_{b,y}(\alpha_{b}) + \sum_{i} {}^{b}\tau_{i,y}(\alpha_{b}, \delta_{i}) + \tau_{n,y} = 0, \tag{5.7}$$

where $\tau_{n,y}$ means the pitch moment from the body, the oil bladders, and the moveable mass as well as the buoyancy of the outer oil bladder. In other words, α_{j} is the combined result of multiple factors, like the center of mass of all parts previously mentioned, the net buoyancy from the outer oil bladder, the whole hydrodynamic coefficients, and the deflection angles of the controllable fins.

From the theoretical analysis in the earlier gliding equilibria, we can conclude that the gliding state can easily be changed by adjusting the controllable fins, including the flippers as well as the fluke. Owing to far from the buoyancy center, the fluke has a relatively larger arm of the force and can afford considerable pitch moment. Therefore, the fluke can play a critical role in pitch attitude control in the gliding process.

5.3 CONTROL METHODS

5.3.1 Heading Control

5.3.1.1 Problem Statement and System Framework

The target of the control system is to make the yaw angle ψ track the reference yaw angle ψ_d by adjusting the deflection angles of the left and right flippers, that is, δ_l and δ_r. The heading dynamics can be formalized as follows:

$$\ddot{\psi} = HB(\Pi + \Gamma_m + \Gamma_o + T_{ext}) + (\Omega_b{}^T \Lambda \Omega_b)/\cos^2\theta \qquad (5.8)$$

with

$$H = \begin{pmatrix} 0 & \dfrac{\sin\phi}{\cos\theta} & \dfrac{\cos\phi}{\cos\theta} \end{pmatrix}$$

$$B = \begin{pmatrix} 0_{3\times3} & I_{3\times3} \end{pmatrix} M^{-1}$$

$$\Lambda = \begin{pmatrix} 0 & \dfrac{1}{2}c\phi c\theta & -\dfrac{1}{2}s\phi c\theta \\ \dfrac{1}{2}c\phi c\theta & 2s\phi c\phi s\theta & s\theta(c^2\phi - s^2\phi) \\ -\dfrac{1}{2}s\phi c\theta & s\theta(c^2\phi - s^2\phi) & -2s\phi c\phi s\theta \end{pmatrix}, \qquad (5.9)$$

where I denotes an identity matrix, c and s in Λ are short for cos and sin, respectively.

In essence, the basic principle of the regulation approach is using differential actions of the flippers to generate asymmetric hydrodynamic forces on the left and right sides for turning. The hydrodynamic forces depend on gliding velocity and attack/sideslip angles. Nevertheless, such gliding states are considerably difficult to measure directly. Up to now, there have been few sensors capable of measuring the attack/sideslip angles for underwater vehicles. Regarding underwater velocity sensing, sensors like Doppler Velocity Logs (DVLs) may be utilized, but their premium prices, large sizes, and low accuracy make them inappropriate for practical application.

Thereby, we design an SMO to estimate the velocities based on data collected from equipped sensors, including a depth sensor and an attitude sensor. With the estimated velocity, we can calculate the attack/sideslip angles. Thereafter, based on the heading dynamics in Equation 5.8, we apply the backstepping methodology to derive a heading control law. Furthermore, in order to convert the controller's instruction to action commands of the flippers, a solver is designed and embedded into the control system. The framework of the heading control system is illustrated in Figure 5.8.

5.3.1.2 SMO-Based Velocity Estimation

Recurring model-based observations is a hopeful approach for dealing with the difficulty of velocity sensing. In terms of the system characteristics, the heading dynamics have a distinct nonlinearity, and the robot's hardware configuration can supply partial state feedback. Exactly, SMO [204][205] is a kind of high-performance state observer suitable for nonlinear uncertain systems with partial state feedback. Hence, we design a SMO for velocity estimation.

The available state data origin from two equipped sensors: an attitude and heading reference system (AHRS) and a depth sensor. The AHRS can provide real-time attitude data in the form of Euler angles and angular velocity. Essentially, the depth sensor acquires the robot's submergence depth d by sensing the water pressure. Thus, the linear velocity along the z-axis of the inertia frame C_g can be derived based on the measured data:

$$^g V_{bz} = \dot{d} - \mathbf{k}^T \cdot {}^g R_b (\Omega_b \times P_d),\qquad(5.10)$$

where d is actually the depth at the point where the depth sensor is installed, denoted by P_d with respect to (w.r.t.) frame C_b.

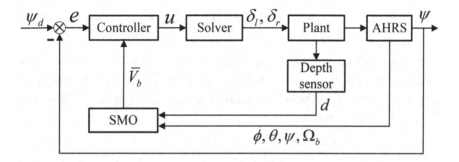

FIGURE 5.8 Framework of the heading control system.

Afterward, we estimate the entire linear velocity vector w.r.t. frame C_g on the basis of ${}^g V_{bz}$. The sliding function s of the SMO is defined as the estimation error of ${}^g V_{bz}$:

$$s = {}^g V_{bz} - {}^g \bar{V}_{bz}, \tag{5.11}$$

where the bar-style superscript indicates the estimated value.

The SMO is designed as follows:

$$
{}^g \dot{\bar{V}}_b = ({}^g R_b \hat{\Omega}_b) \bar{V}_b + {}^g R_b \dot{\bar{V}}_b + \begin{pmatrix} k_x \mathrm{sgn}(s) \\ k_y \mathrm{sgn}(s) \\ k_z \mathrm{sgn}(s) \end{pmatrix}, \tag{5.12}
$$

$$
\dot{\bar{V}}_b = A \left[\Pi(\bar{V}_b) + T_{ext}(\bar{V}_b) + \Gamma_m + \Gamma_o \right]
$$

where $K = \mathrm{diag}(k_x, k_y, k_z)$ is the gain matrix of the SMO. $T_{ext}(\cdot)$ and $\Pi(\cdot)$, are functions of $\bar{V}_b$ rather than V_b here. $A = \begin{pmatrix} I_{3\times3} & 0_{3\times3} \end{pmatrix} \mathrm{M}^{-1}$. $\mathrm{sgn}(\cdot)$ indicates the sign function.

The dynamics of estimation error take the following forms:

$$
{}^g \dot{\tilde{V}}_b = ({}^g R_b \hat{\Omega}_b) \tilde{V}_b + {}^g R_b \dot{\tilde{V}}_b - \begin{pmatrix} k_x \mathrm{sgn}(s) \\ k_y \mathrm{sgn}(s) \\ k_z \mathrm{sgn}(s) \end{pmatrix}, \tag{5.13}
$$

$$
\dot{\tilde{V}}_b = A \left[\Pi(V_b) - \Pi(\bar{V}_b) + T_{ext}(V_b) - T_{ext}(\bar{V}_b) \right]
$$

where ${}^g \tilde{V}_b = {}^g V_b - {}^g \bar{V}_b$, and the tilde-style superscripts represent the estimation errors. The sliding condition for the SMO is $s\dot{s} < 0$, which is equivalent to

$$
k_z > \left| \mathrm{k}^T \left(({}^g R_b \hat{\Omega}_b) \tilde{V}_b + {}^g R_b \dot{\tilde{V}}_b \right) \right|. \tag{5.14}
$$

Assuming that s reaches the sliding surface, that is, $\dot{s} = 0$, we apply Philippov's theory of equivalent dynamics [21] to Equation 5.10 and thereafter locally linearize the equations around ${}^g \bar{V}_b$. As a result, new forms of ${}^g \dot{\tilde{V}}_{bx}$ and ${}^g \dot{\tilde{V}}_{by}$ are derived

$$
\begin{pmatrix} {}^g \dot{\tilde{V}}_{bx} \\ {}^g \dot{\tilde{V}}_{by} \end{pmatrix} = Q \begin{pmatrix} {}^g \tilde{V}_{bx} \\ {}^g \tilde{V}_{by} \end{pmatrix} \tag{5.15}
$$

with

$$Q = \begin{pmatrix} \dfrac{\partial f_x}{\partial^g V_{bx}} - \dfrac{k_x}{k_z}\dfrac{\partial f_z}{\partial^g V_{bx}} & \dfrac{\partial f_x}{\partial^g V_{by}} - \dfrac{k_x}{k_z}\dfrac{\partial f_z}{\partial^g V_{by}} \\[3mm] \dfrac{\partial f_y}{\partial^g V_{bx}} - \dfrac{k_y}{k_z}\dfrac{\partial f_z}{\partial^g V_{bx}} & \dfrac{\partial f_y}{\partial^g V_{by}} - \dfrac{k_y}{k_z}\dfrac{\partial f_z}{\partial^g V_{by}} \end{pmatrix} \tag{5.16}$$

$$\left(f_x, f_y, f_z\right) = \left(i^T \cdot^g \dot{V}_b,\ j^T \cdot^g \dot{V}_b,\ k^T \cdot^g \dot{V}_b\right).$$

It is revealed from Equation 5.15 that appropriate k_x and k_y can be selected to make Q a Hurwitz matrix in order to ensure the convergency of $^g\tilde{V}_{bx}$ and $^g\tilde{V}_{by}$.

Actually, to eliminate undesirable chattering effect in practice, the sign function is replaced by a saturation function defined as follows:

$$\text{sat}(s,\varepsilon) = \begin{cases} s/\varepsilon & \text{if } |s/\varepsilon| \le 1 \\ \text{sgn}(s/\varepsilon) & \text{if } |s/\varepsilon| > 1 \end{cases}, \tag{5.17}$$

where ε is a positive value and indicates the boundary layer thickness.

5.3.1.3 Controller Design

Based on the heading dynamics, we employ the backstepping methodology [206]–[208] to design a heading controller. It is owing to that backstepping is a recursive design approach suitable for cascade systems. In the course of applying backstepping, not only are feedback control laws deduced, but also stability is guaranteed by establishing associated Lyapunov functions. The procedures are as follows:

Step 1: Define the heading error by $e = \psi - \psi_d$. The first backstepping state variable is chosen as z_1

$$\begin{aligned} z_1 &= e \\ \dot{z}_1 &= \dot{\psi} - \dot{\psi}_d = \gamma - \dot{\psi}_d. \end{aligned} \tag{5.18}$$

Then, we regard γ as virtual control input

$$\gamma = \xi + z_2, \tag{5.19}$$

where z_2 is a new state variable to be interpreted later and ξ is a stabilizing function, yielding

$$\dot{z}_1 = \xi + z_2 - \dot{\psi}_d. \tag{5.20}$$

Afterward, to facilitate $z_1 \to 0$, we can design the stabilizing function ξ as follows:

$$\xi = \dot{\psi}_d - c_1 z_1, \tag{5.21}$$

where $c_1 > 0$ denotes a feedback gain, yielding

$$\dot{z}_1 = -c_1 z_1 + z_2. \tag{5.22}$$

Furthermore, we define a Lyapunov function V_1 for z_1:

$$V_1 = \frac{1}{2} z_1^2$$
$$\dot{V}_1 = z_1 \dot{z}_1 = -c_1 z_1^2 + z_1 z_2. \tag{5.23}$$

Step 2: In this step, we stabilize the dynamics of z_2, which can be formalized as follows:

$$\dot{z}_2 = \dot{\gamma} - \dot{\xi} = H \left[B(\Pi + \Gamma_m + \Gamma_o + T' + T_p) \right] + \Omega_b^T \Lambda \Omega_b - \dot{\xi}, \tag{5.24}$$

where the force exerted by the flippers (denoted by T_p) is separated from the total external force T_{ext}, and $T' = T_{ext} - T_p$.

We define the second Lyapunov function as follows:

$$V_2 = V_1 + \frac{1}{2} z_2^2, \tag{5.25}$$

and derive its derivative

$$\dot{V}_2 = \dot{V}_1 + z_2 \dot{z}_2 = -c_1 z_1^2 + z_2 [z_1 + \Omega_b^T \Lambda \Omega_b - \dot{\xi} \\ + HB(\Pi + \Gamma_m + \Gamma_o + T' + T_p)] \tag{5.26}$$

Design a control law u as follows:

$$u = HBT_p = -HB(\Pi + \Gamma_m + \Gamma_o + T') \\ - \Omega_b^T \Lambda \Omega_b + \dot{\xi} - z_1 - c_2 z_2, \tag{5.27}$$

where $c_2 > 0$ is another design parameter, yielding

$$\dot{V}_2 = -c_1 z_1^2 - c_2 z_2^2. \tag{5.28}$$

Apparently, $\dot{V}_2 < 0, \forall z_1 \neq 0, z_2 \neq 0$. Hence, by applying control law, the heading control system is asymptotically stable.

The control law can be rearranged as follows:

$$u = -HB(\Pi + \Gamma_m + \Gamma_o + T') - \Omega_b^T \Lambda \Omega_b + \ddot{\psi}_d - h_1 \dot{z}_1 - h_2 z_1, \tag{5.29}$$

where $h_1 = c_1 + c_2 \, (h_1 > 0)$ and $h_2 = c_1 c_2 + 1 \, (h_2 > 1)$. In essence, the physical meaning of u is the desired yaw acceleration exerted by the flippers.

5.3.1.4 Solver for Flipper Action Commands

It is necessary to design a solver to resolve the controller's instructions to action commands of the flippers. Nevertheless, we are confronted with four obstacles:

1. *Control allocation*: The two flippers are redundant control effectors for heading regulation, while the controller's outputs are with reduced dimension. Thereby, the solver needs to map a reduced dimension instruction to action commands of the two flippers. Such a problem is referred to as control allocation [209].

2. *Nonlinearity of hydrodynamics*: The relation between deflection angles and resultant yaw moments is nonlinear (see Figure 5.9), especially when the flippers rotate in a wide angular range. How to solve deflection angles of the flippers for desired yaw moment is a problem.

3. *Jitter of deflection angles*: The nonlinearity of hydrodynamics and the underdetermined property of the control allocation problem will lead to multiple solutions when solving for the deflection angles. As shown in Figure 5.9, the same yaw moment may correspond to different deflection angles, and their interval may be large. Thereby, a jitter of deflection angles will arise during continuous control, which leads to considerable but undesirable energy consumption.

4. *Constraints of exit status*: At the end of heading regulation, we expect that not only will the heading angle reach the target value, but also the attendant motion will vanish. Specifically, turning is usually accompanied by certain roll and sideslip angles, owing to the system's coupling.

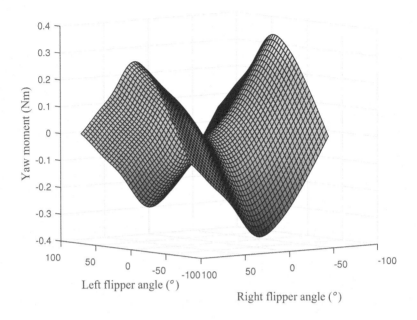

FIGURE 5.9 Yaw moment exerted by a pair of flippers when V_b= 0.4 m/s, V_{by} = 0, α_b = 20°, and $\Omega_b = 0_{3\times1}$.

These side effects may stay after heading regulation ends, which will increase drag forces and thereby result in unnecessary energy loss. Hence, the solver should also meet the constraints of the exit status.

The four preceding problems are synthesized into the following optimization problem:

$$\underset{\delta_l,\delta_r}{\arg\min}[u - HBT_p(\delta_l,\delta_r,\bar{V}_b,\Omega_b)]^2 + \lambda_1(\delta_l - \delta_r)^2 + \lambda_2[(\delta_l - \delta_l')^2 + (\delta_r - \delta_r')^2]$$

$$\delta_l \in \left[\max(-\frac{\pi}{2}, -\frac{\pi}{2}+\bar{\alpha}_{bl}), \min(\frac{\pi}{2}, \frac{\pi}{2}+\bar{\alpha}_{bl}) \right]$$

$$\delta_r \in \left[\max(-\frac{\pi}{2}, -\frac{\pi}{2}+\bar{\alpha}_{br}), \min(\frac{\pi}{2}, \frac{\pi}{2}+\bar{\alpha}_{br}) \right],$$

(5.30)

where λ_1 and λ_2 are coefficients. δ_l' and δ_r' denote the flippers' deflection angles at the previous control period. $T_p(\cdot)$ is expressed as the function of $\delta_l, \delta_r, \bar{V}_b$, and Ω_b.

The first term in the objective function is oriented to the problem of control allocation. The second term aims to eliminate the side effects caused by turning motion through gradually guiding δ_l and δ_r to a same value when the heading error is closed to zero. It is according to that symmetrical forces will not support nonzero roll angles and sideslip angles. Additionally, the third term punishes jitter of the deflection angles to smooth the flippers' actions. The constraints of δ_l and δ_r limit the flipper's AoAs to $[-\frac{\pi}{2}, \frac{\pi}{2}]$. Considering the nonlinearity, we apply the particle swarm optimization (PSO) algorithm to search optimal deflection angles.

To decrease possible interference to the convergence process of heading error, the values assigned to λ_1 and λ_2 should be properly small.

5.3.2 Pitch Control

This section presents a pitch control method to regulate the pitch attitude of the robotic dolphin in the gliding process by adjusting the controllable fluke. Before designing the pitch controller for the robotic dolphin, we first formalize its pitch dynamics

$$\ddot{\theta} = LB(\Pi + \Gamma_m + \Gamma_o + T'_{ext}) + \Omega_b{}^T W \Omega_b + u \tag{5.31}$$

with

$$
L = \begin{bmatrix} 0 & \cos\phi & -\sin\phi \end{bmatrix}
$$

$$
B = \begin{pmatrix} 0_{3\times3} & I_{3\times3} \end{pmatrix} M^{-1}
$$

$$
W = - \begin{pmatrix} 0 & \frac{1}{2}s\phi & \frac{1}{2}c\phi \\ \frac{1}{2}s\phi & s^2\phi t\theta & s\phi c\phi t\theta \\ \frac{1}{2}c\phi & s\phi c\phi t\theta & c^2\phi t\theta \end{pmatrix}, \tag{5.32}
$$

where T_f and T'_{ext} separately denote the extern forces from the fluke and other parts, and $T'_{ext} = T_{ext} - T_f$. I denotes an identity matrix. c, s, and t in W are short for cos, sin, and tan, respectively. $u = LBT_f$ denotes the control valuable, which can be regulated through adjusting the deflection angle of the fluke δ_f.

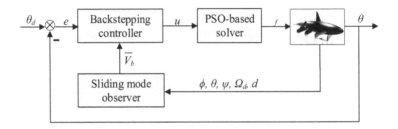

FIGURE 5.10   Schematic of the proposed pitch control system.

Figure 5.10 illustrates the scheme of the proposed pitch control system. The control system consists of three parts: a backstepping controller, a PSO-based solver, and a sliding mode observer. Specifically, the backstepping controller is constructed to derive a pitch control law based on the pitch dynamic. the PSO-based solver is to convert the controller's instruction to the fluke's action, and the sliding observer is built to estimate the gliding velocities via the sensor data to further calculate the attack and sideslip angles for the backstepping controller.

5.3.2.1 Backstepping Controller

Here, we employ the backstepping technique to design the pitch controller. Given a desired pitch angle θ_d, the objective is to design control laws with the backstepping technique to make sure the gliding robotic dolphin track θ_d with tiny errors by regulating the fluke.

Here we directly provide the control law u based on the backstepping approach for the robotic dolphin.

$$u = -LB(\Pi + \Gamma_m + \Gamma_o + T_{ext}) - \Omega_b{}^T W \Omega_b + \ddot{\theta}_d - (c_1 + c_2)\dot{z}_1 - (c_1 c_2 + 1)z_1, \quad (5.33)$$

where $c_1 > 0$ and $c_2 > 0$ are feedback gains of the controller. $z_1 = \theta - \theta_d$ and $\dot{z}_2 = \dot{\theta} - \dot{\theta}_d$ denote the pitch error and its differential, respectively. In fact, the physical meaning of the control law u is the desired pitch acceleration from the fluke.

5.3.2.2 PSO-Based Solver

As mentioned previously, the designed pitch controller outputs desired pitch acceleration exerted by the fluke. Thus, it is necessary to resolve the controller's instructions to action commands of the fluke. In fact, the hydrodynamics of the robot have severe nonlinearity, implying that the relationship between deflection angle and resultant pitch moment is nonlinear, especially when the fluke rotates in a wide angular range. Therefore, the deflection angle of the fluke has no expression based on the dynamics. The following problem is to solve the deflection angle of the fluke for desired pitch moment. Here, we construct an optimization problem as follows:

$$\begin{cases} \underset{\delta_f}{\arg\min}[[u - LBT_f(\delta_f, \overline{V}_b, \Omega_b)]^2 + \lambda_f(\delta_f - \delta_f')^2 \\ \delta_f \in \left[\max(-\frac{\pi}{3}, -\frac{\pi}{2} + \overline{\alpha}_{bf}), \min(\frac{\pi}{2}, \frac{\pi}{2} + \overline{\alpha}_{bf}) \right] \end{cases} \quad (5.34)$$

with

$$\overline{\alpha}_{bf} = \arctan(\frac{^b\overline{V}_{fz}}{^b\overline{V}_{fx}}), \quad (5.35)$$

where λ_f is a coefficient. δ_f' is the fluke's deflection angles over the previous control period. $T_f(\cdot)$ is expressed as the function of δ_f, $\overline{V}_b$, and Ω_b. In detail, the first term in the objective function aims to the problem of control allocation. The second term punishes the jitter of the deflection angle to smooth the fluke's actions. Here, the PSO algorithm [211] is applied to search optimal deflection angles. Note that λ_f should be assigned a properly small value to decrease possible interference to the convergence process of pitch error.

5.4 SIMULATION RESULTS AND ANALYSIS

5.4.1 Heading Simulation

5.4.1.1 Results of Velocity Estimation

A typical 3D gliding case, that is, spiraling motion, is utilized to validate the performance of the SMO-based velocity estimation method. At the initial time, the gliding robotic dolphin is stationary with zero-net buoyancy and zero-attitude angles. At $t = 1$ s, the pump begins to rotate at a speed of $v_o = 50$ rps (revolutions per second) to suck the oil from the external bladder to the inner one until the external bladder is empty. Meanwhile, the movable mass moves forward at 3 mm/s for a distance of 3.9 cm. Additionally, the left flipper stays at $\delta_l = 15°$ to generate turning moment. Consequently, the robot will spiral downward. The gain matrix and the initial state of the SMO are $K = \text{diag}\{50, 50, 50\}$ and $^g\overline{V}_b = [0.1, 0.1, 0.1]^T$. Figure 5.11 shows the result of velocity estimation. It is shown that the

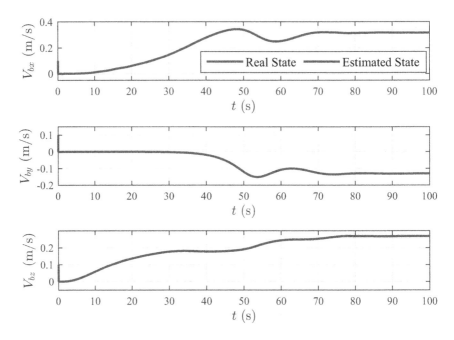

FIGURE 5.11   Velocity estimation without parameter uncertainty and external disturbance.

estimated states catch up with the real ones rapidly and thereafter stay overlapped with them.

Furthermore, both parameter uncertainty and external disturbance are introduced to verify the robustness of the SMO. The parameter uncertainty is quantized by a relative model error that is superposed to the formula of $\dot{V}_b$. For instance, 20% model error indicates $0.2\dot{V}_b$. The external disturbance is modeled by a time-varying function $\chi = q\sin(t)$, which is also superposed to $\dot{V}_b$. Figure 5.12 displays the estimation result under 50% model error and disturbance with an amplitude of $q = 0.01$ m^2/s. It is noticed that obvious velocity fluctuations appear since the disturbance changes the robot's acceleration directly. Although suffering from a strong model error and disturbance, the estimated velocities keep close to the real ones with small deviations. In addition, the SMO is also tested under different parameter uncertainty and external disturbance. The estimation errors are plotted in Figure 5.13. With the distinct increase of parameter uncertainty and disturbance, the estimation errors rise but still fluctuate in an accepted range, which demonstrates the robustness of the SMO.

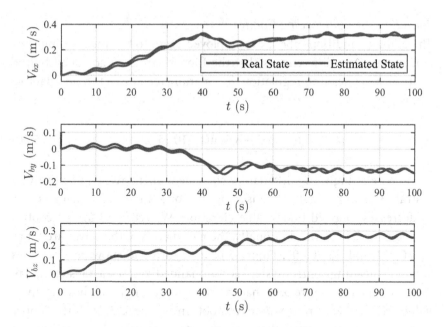

FIGURE 5.12   Velocity estimation under 50% model error and external disturbing term $\chi = 0.01\sin(t)$.

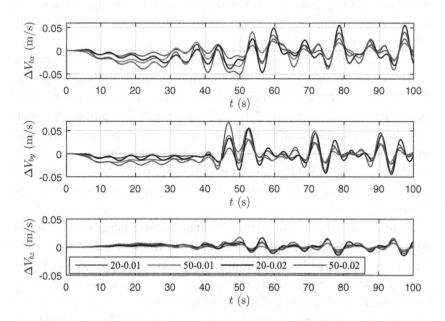

FIGURE 5.13   Errors of velocity estimation under $p\%$ model error and external disturbing term $\chi = q\sin(t)$, which are marked by the format of "p-q."

5.4.1.2 Results of Heading Control

The gliding conditions above are employed again to test the heading control system. The difference is that the flippers are now governed by the controller for heading regulation. The parameter h_1 is set in an adaptive format

$$h_1 = 10.5 + 8.5\text{sat}(e - 10^{\circ}, 5^{\circ}). \tag{5.36}$$

When e is high, the differential coefficient h_1 increases adaptively to restrain possible overshoot caused by the robot's inertia. When e is low, h_1 decreases to avoid undesirable vibration. We set $h_2 = 1.5$. The control period is set as 0.2 s. Additionally, the parameters of the solver are set as $\lambda_1 = 0.002$ and $\lambda_2 = 0.003$. Figure 5.14 to Figure 5.16 show part of states when the heading control system is commanded to track a 45° step signal. It is shown from Figure 5.14 that the robot's yaw angle reaches the given value precisely with nearly no overshoot. In Figure 5.15, both the flipper angles jump to the values that can generate a maximum yaw moment at $t = 40$ s and jump again reversely at $t = 44$ s to decelerate the turning speed to avoid overshoot. Apparently, from $t = 67$ s, both the flippers are compelled gradually to a same angle by the penalty term on $|\delta_l - \delta_r|$. Thereby, the roll angle in Figure 5.14, V_{by} in Figure 5.16, and the sideslip angle β_b returns to zero at the same time.

To validate the robustness of the combination of the SMO and the controller, we introduce 50% model error and external disturbing term $\chi = 0.01\sin(t)$ at $t = 40$s as in Figure 5.12. Some states and the flippers' angles are shown in Figures 5.17 and Figure 5.18. In spite of the unfavorable factors, the heading error still keep closed to 0 and fluctuates within $\pm 0.4^{\circ}$. The roll angle and V_{by} also undulate around 0 in considerably tiny ranges. When the heading error is small, both the flippers rotate around a same angle to generate a yaw moment that resists the disturbance.

Moreover, we also investigate the effect of the parameter λ_1 in the solver on the exit status of heading control, as listed in Table 5.2. When $\lambda_1 = 0$, there is completely no limit on the difference between δ_l and δ_r. As a result, although the yaw angle reaches the target value, the two flippers are stuck in two angles with a vast difference, which generates a roll moment balanced by the net buoyancy and thereby a roll angle of 2.6°. Additionally, the consequent lateral force component of the flippers prompts the body

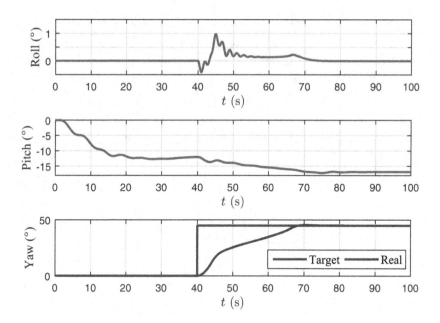

FIGURE 5.14 Attitude curves when the heading control system is commanded to track a 45° step signal.

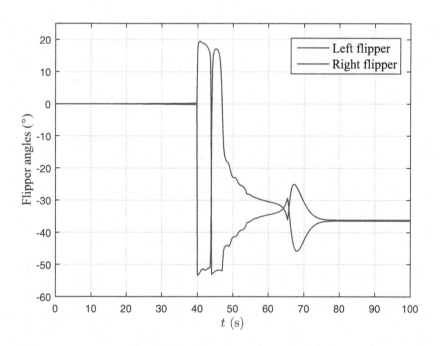

FIGURE 5.15 Deflection angles of the flippers when the heading control system is commanded to track a 45° step signal.

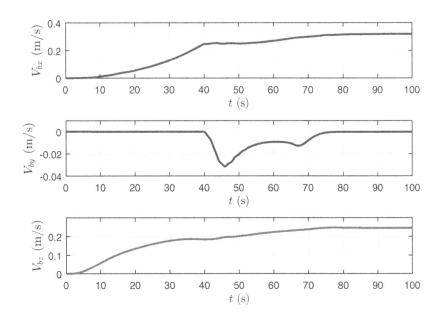

FIGURE 5.16   Velocity curves when the heading control system is commanded to track a 45° step signal.

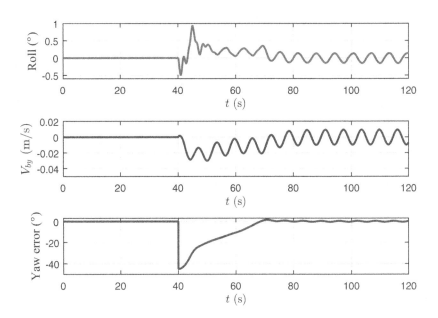

FIGURE 5.17   States of heading control under 50% model error and external disturbing term $\chi = 0.01\sin(t)$.

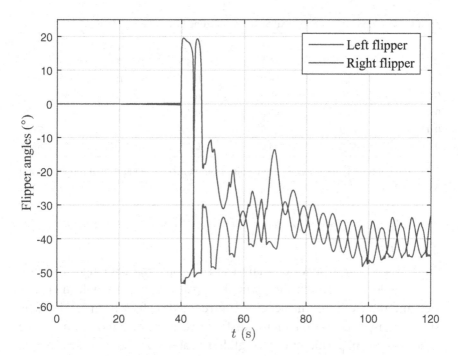

FIGURE 5.18 Deflection angles of the flippers during heading control under 50% model error and external disturbing term $\chi = 0.01\sin(t)$.

to sideslip, which generates a lateral force in turn to counteract the flippers' effect. Certainly, the roll angle also contributes to the body's lateral force by rotating the body's lift-and-drag forces out of the vertical plane.

Essentially, the robot does not glide exactly toward the direction that its body points to, in this case. Table 5.2 reveals that the penalty term on $\left|\delta_l - \delta_r\right|$ in the solver can eliminate such side effects by guiding the flippers gradually to a same angle. If the flippers are simply adjusted to a same angle after the heading error converges, the achieved balance of yaw and roll moment will be destroyed, and thereby, the robot will instantly departure from the target heading direction. To some extent, the increase of λ_1 can accelerate the convergence of the flippers' angle difference. However, as the cases under $\lambda_1 = 0.008$ and $\lambda_1 = 0.01$, a too-large λ_1 affects the heading regulation and results in overshoot, which decelerates the convergency of the heading error as well as the flippers' angle difference on the contrary.

TABLE 5.2   Effect of λ_1 on the Exit Status of Heading Control

λ_1	$\delta_l(\circ)$	$\delta_r(\circ)$	$\phi(\circ)$	$\beta_h(\circ)$	$t_e(\circ)$
0	−31.6	44.1	2.6	2.3	–
0.002	−36.4	−36.2	0	0	78.8
0.004	−43.7	−43.7	0	0	72.8
0.005	−43.8	−43.8	0	0	72.6
0.008	−43.9	−43.9	0	0	79.2
0.01	−44.0	−44.0	0	0	85.0

Note: The notation t_e indicates the time point from which the left and right flippers stay at a same angle.

5.4.1.3 Discussion

To deal with the absence of analytical solutions, the PSO algorithm has been employed to search the optimal flipper angles. Although the precision and robustness of the heading control method have been validated by simulations, it seems difficult to satisfy the realtime demand. Actually, the heading control period of typical gliders is about 2 ~ 10 s, to save energy. The average elapsed time of the solver in MATLAB® is 1.2 s, when the PSO has a population size of 40 and the angular search space is continuous. Parallel implementation of the PSO by C/C++ language and discretizing the search space will dramatically decrease the elapsed time. In addition, fuzzy strategy is an alternative to approximate the solver and is more suitable for practical applications, at the cost of precision and convergence speed to some extent.

The penalty term on $|\delta_l - \delta_r|$ in the solver can lead the two flippers to a same angle. But the consequent angle is not necessarily beneficial to the gliding motion, since it has not been taken into consideration in the solver. There are two solutions to the problem. One is to add another penalty term in the solver, which can simultaneously optimize the gliding performance during heading regulation. The other is simply adjusting the flippers to a desired angle at the same time after they converge at the same angle. It is feasible since the forces are now symmetrical on the left and the right. Besides, Figure 5.18 shows that the periodic disturbance results in an undulation of the flippers and thereby energy consumption. Adding a dead band to the controller is an option to reduce excessive actions by the flippers at the expense of accepted precision loss.

5.4.2 Pitch Simulation

In this section, typical simulations in MATLAB/Simulink are conducted to verify the effectiveness of the proposed pitch control method for the gliding robotic dolphin. The hydrodynamic coefficients, like lift coefficient, drag coefficient, and moment coefficient are obtained via the CFD simulation.

A case of pitch control is simulated, in which the reference pitch angle is set as −25° in the downward gliding phase, and 25° in upward gliding phase, respectively. Before gliding, the robotic dolphin is stationary with zero net buoyancy and zero attitude angle. At $t = 1$ s, the buoyancy-driven system starts to work, and the pump rotates at a speed of 50 rps (revolutions per second) to suck the oil from the external bladder to the inner one, until that the external bladder is empty. Meanwhile, the movable mass moves forward at 3 mm/s for a distance of 3.9 cm. Consequently, the robotic dolphin glides downward. Figure 5.19 illustrates the whole gliding process. In the first 90 s, the pitch controller does not work, and the robotic dolphin glides downward freely with a zero-deflection angle of the fluke, as shown in Figure 5.19a. When $t = 90$ s, the pitch controller begins to function. Simultaneously, the robotic dolphin regulates its fluke to generate the expected moment and the pitch angle begins to track the reference value −25° quickly. After about 20 s, the pitch angle reaches a steady value at −25° with a deflecting fluke of −6.7°. When $t = 220$ s, an unsteady gliding state appears, because the robotic dolphin begins to push off the oil to the extern oil bladder for gliding upward. At this time, the pitch controller is off. Under the upward-force moments, the pitch angle starts getting smaller and finally becomes a positive one for gliding upward. When $t = 350$ s, the pitch controller begins to work again, and the pitch angle in the upward gliding phase quickly tracks the desired one 25°. The fluke gains a relatively large deflection angle, about 20°. Note that the flippers do not work in the whole process and always keep zero-deflection angles. Besides, we also simulate the pitch angle curve in gliding with the same conditions except no pitch controller. As shown in Figure 5.19a, the robotic dolphin finally obtains a static pitch angle of about 20° when gliding freely. This illustrates that the proposed control method is effective, and the fluke can be employed to realize pitch control. Figure 5.19c plots the trajectory of the whole gliding motion, in which the robotic dolphin realizes a "V" gliding motion in the vertical plane. As for the PSO-based solver, the number of particle swarm is set as 40, and the search space is continuous. Through counting, it takes about 1.2 s in MATLAB.

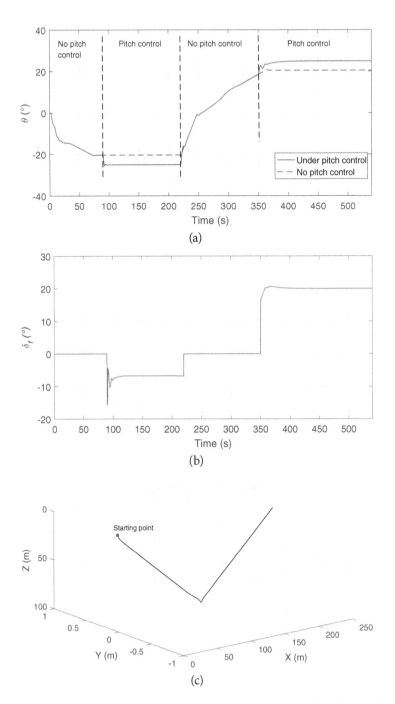

FIGURE 5.19   Simulation results in tracking a desired pitch angle (gliding down-ward: –25°; gliding upward: 25°). (a) Pitch angle curves. (b) The deflection angle of the fluke. (c) Trajectory.

5.5 EXPERIMENTAL RESULTS AND ANALYSIS

5.5.1 Heading Experiment

For the embedded computing platform carried by the gliding robotic dolphin, it is almost impossible to completely implement the algorithm. A feasible measure is to simplify the pectoral fin deflection angle solver at the cost of a certain control effect and only adjust the heading through the unilateral pectoral fin. In this way, only a single pectoral fin angle needs to be solved to satisfy the controller commands. In this chapter, the grid search method is used to solve the pectoral fin angle that is closest to the control command, and the smooth term of the pectoral fin motion and the constraint term for the difference in the two pectoral fin motions are ignored, namely, $\lambda_1 = 0$ and $\lambda_2 = 0$. In this chapter, the control method is verified by underwater experiments. The search step of the pectoral fin angle is set to 5°, and the dead zone of heading adjustment is set to ±2°. The gliding robotic dolphin absorbs oil to dive into the water.

Figure 5.20 illustrates the snapshot sequence of gliding motion with heading control. The speed of the oil pump is 2000 rpm. At the beginning, the target heading is set as the direction of pool's central axis, and the error of the yaw angle is 30°. When $t = 70$ s, the gliding robotic dolphin dives to the bottom of the pool, and for the slower adjustment of buoyancy, it starts to float at $t = 125$ s. when $t = 169$ s, the robotic gliding robotic dolphin reaches the other end of the pool. As can be seen from the figure, when the

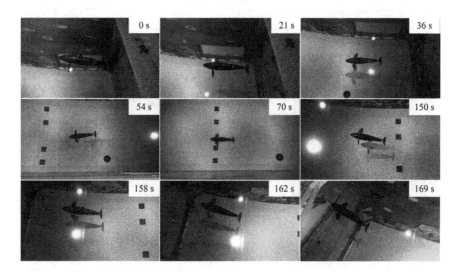

FIGURE 5.20   The sequence of the gliding motion heading control experiment.

robotic dolphin deviates from the target direction, the controller corrects the heading deviation through unilateral pectoral fin action. During the gliding motion, the heading of the gliding dolphins remained basically parallel to the side wall of the pool, and this experimental example intuitively demonstrated the effect of the proposed control method.

Figure 5.21 presents two experimental results of heading control, the toggle depth is set as 6.5 m. Before $t = 40$ s, the gliding robotic dolphin is

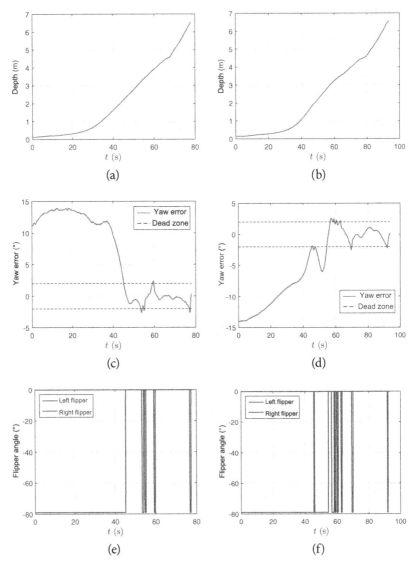

FIGURE 5.21   The experimental data of heading control. (a) Depth 1. (b) Depth 2. (c) Yaw error 1. (d) Yaw error 2. (e) Deflection angles of the flippers 1. (f) Deflection angles of the flippers 2.

in the acceleration phase, the control torque generated by the pectoral fins is weak, and the heading adjustment is slightly slower. After entering the steady state, under the adjustment action of the controller, the yaw error can quickly converge and maintain within the adjustment dead zone. Note that when the heading is adjusted, the pectoral fin deflection angle is at the limit. There are two reasons for this "extreme" action of the pectoral fin: (1) For the limited size of the experimental pool, we hope that the gliding robotic dolphin can quickly converge the heading error before touching the pool boundary. Besides, the set control parameters are too large, so the control effect is violent. (2) When the preceding control method is implemented on the hardware platform, the smoothing term for the pectoral fin movement is ignored. In underwater environments with greater depth and breadth, more gentle control effects may be obtained by setting appropriate parameters and smoothing the pectoral fin movements, which will be further tested. The conducted experiments have verified the effectiveness of the proposed heading control method.

5.5.2 Pitch Experiment

The penalty term on $|\delta_l - \delta_r|$ in the solver can lead the two flippers to the same angle. In order to explore the effect of the controllable fluke for the pitch attitude of the gliding robotic dolphin, extensive experiments were conducted in a diving pool with a dimension of 16 m long, 6 m wide, and 7.5 m deep. The robotic dolphin was released in one side of the pool and required to glide to the other side. A video camera was employed to record the whole process. Meanwhile, the onboard sensors collected the gliding data such as diving depth and attitudes in real time for further analysis.

5.5.2.1 Testing of Gliding Motion with Different Fluke Deflection Angles

First, the gliding performance in different deflection angles of the fluke was tested. As is demonstrated, the fluke can generate considerable pitch moment even if its area is not very large, because of a relatively larger arm of force. In order to quickly enter the steady state, the robotic dolphin was released after the oil was pumped off in the following experiments. Meanwhile, the fluke was set to different deflection angles from $-80°$ to $80°$. Figure 5.22 shows a snapshot sequence of a downward gliding motion. Besides, the gliding depths and pitch angles of the robotic dolphin with different fluke angles are plotted in Figure 5.23, and the relevant statistical data are listed in Table 5.3 and Table 5.4. When $\delta_f = 30°, 60°, 80°$, the AoA α_f

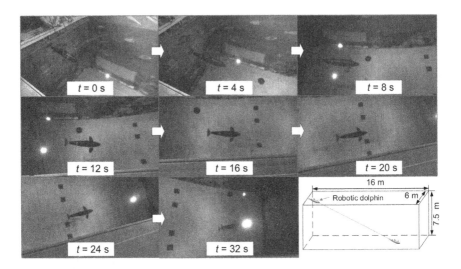

FIGURE 5.22   Snapshot sequence of a downward gliding motion with a deflecting fluke.

becomes −1; thus, the lift force from the fluke is toward a down direction. As a result, the resultant force of lift and drag pointing to the body oblique below produces an upward force moment, and the pitch angle θ increases. Rather, the pitch angle θ will decrease when $\delta_f = -30°, -60°, -80°$. Actually, we can directly compare θ when $\alpha_f > 0$ with that when $\alpha_f < 0$, since it is easy to get that $C'_{j,l}$ has opposite signs and $C'_{j,d}$ has the same sign in these two situations. Besides, it can be observed that the AoA α_b increases with $|\delta_f|$, as shown in Table 5.3. As for the AoA α_f, when $\delta_f \in [0°, 30°]$, α_f converts from negative to positive. Beyond this range, $|\alpha_f|$ increases with $|\delta_f|$. However, due to the complex nonlinear hydrodynamic forces, the relationship between θ and δ_f is fairly non-intuitive.

As shown in Table 5.3, the gliding angle $|\gamma|$ increases with $|\delta_f|$, indicating the gliding trajectory becomes steeper and steeper. The main reason is that with the increase of $|\delta_f|$, $|\alpha_b|$ increases and $|\alpha_f|$ also increases in a certain range. Under a large AoA, the lift does become little, but the drag becomes larger. As a result, the whole lift-to-drag ratio decreases. It is not difficult to find that $|\gamma|$ will become larger when the lift-to-drag ratio decreases. Furthermore, when $\delta_f = -60°, -80°$, the body obtains the largest drag, and the fluke is almost perpendicular to the flow and generates the largest drag. Therefore, $|\gamma|$ becomes the largest. Besides, the gliding

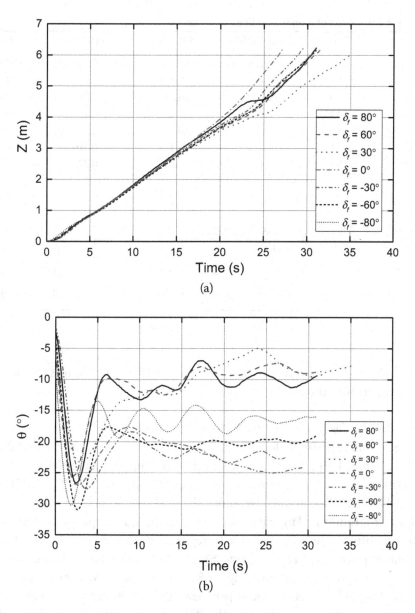

FIGURE 5.23   Plots of the gliding performance of the robotic dolphin with different fluke deflection angles. (a) Gliding depth. (b) Pitch angle.

velocity decreases with $|\delta_f|$ in Table 5.3. This situation can be analyzed as follows. Given an m_0, $\|{}^g V_b\|$ is decided by the sum of the squares of the lift and drag coefficients. It can be inferred that under a certain large AoA, the stall occurs, in which the lift is little, but the drag sharply increases.

TABLE 5.3   Effect of δ_f on the Gliding Performance.

$\delta_f(\circ)$	X(m)	^gV_x (m/s)	^gV_z (m/s)	^gV (m/s)	$\gamma(\circ)$
−80	4.4	0.126	0.217	0.251	−59.9
−60	5.8	0.169	0.220	0.277	−52.5
−30	8.3	0.255	0.232	0.345	−42.3
0	10.2	0.340	0.250	0.422	−36.3
30	12.4	0.315	0.191	0.368	−31.3
60	7.6	0.221	0.218	0.310	−44.6
80	5.5	0.161	0.220	0.273	−53.8

TABLE 5.4   Effect of δ_f on the AoA α_b, and α_f

$\delta_f(\circ)$	−80	−60	−30	0	30	60	80
$\alpha_b(\circ)$	44	33	18	13	23	36	44
$\alpha_f(\circ)$	124	93	48	13	−7	−24	−36

Therefore, $\|^g V_b\|$ decreases with $|\delta_f|$. So does $\|^g V_{bx}\|$. It agrees well with the experimental data.

5.5.2.2 Testing of Pitch Control Based on the Fluke

The second experiment focused on pitch control based on the controllable fluke. Considering the difficulty in real-time implementation of the PSO optimization in the embedded controller (MCU: STM32F407, 168MHz, 1M Flash, 192Kb SRAM), we adopt the grid search method through C/C++ programming language. In particular, the search space of the fluke's angle is discretized, and the search step is set as $\pm 2^\circ$. Meanwhile, the dead zone of pitch regulation is set as $\pm 2^\circ$. In this way, the PSO-based solver can almost work in real time. Figure 5.24 shows the relevant experimental snapshot and results. In the experiment, the referenced pitch angle was set to -12°, and the pitch controller did not work until the diving depth was beyond 0.3 m, where the robotic dolphin fully submerged. At first, the robot kept still in the surface of the water and started to glide downward after the pump sucked the oil at the speed of 40 rps. About 15 s later, the diving depth was beyond 0.3 m, and the pitch controller began to work, as shown in Figures 5.24b–c. Consequently, the error of pitch angle converges gradually under the effect of the controllable fluke. The experimental results validate the effectiveness of the proposed pitch control system

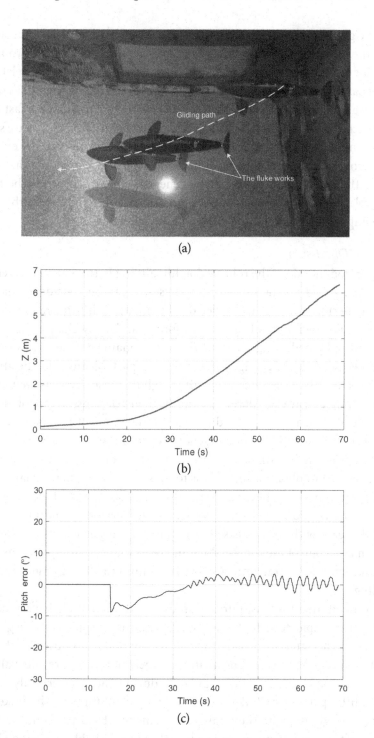

FIGURE 5.24 Illustration of pitch control for the developed robotic dolphin. (a) Experimental snapshot. (b) Plot of the diving depth. (c) Plot of the pitch-angle error.

for the gliding motion based on the fluke. More careful inspection shows that the error has slight undulation in a range of $\pm 2°$. There are two main reasons which contribute to this slight undulation. One is the relatively large parameters in the pitch controller. Because of the diving pool's small size, we set some relatively large control parameters to pursue a fast convergence. These parameters easily lead to a strong reaction and cause some pitch undulation. The other is due to the heading controller in the embedded programs. In fact, the heading controller also worked at the same time. Therefore, the flippers maybe generate some disturbs for the pitch control, and the fluke needs to be regulated to cancel these disturbs. This also imports some slight undulation in the pitch direction.

5.5.2.3 Discussion

The hydrodynamic of the robotic dolphin gliding in the water is extremely complex. Based on the preceding analyses, it can be concluded that the gliding performance is largely decided by the whole lift-to-drag ratio. The flatten fluke, owing to a large arm of force, can largely change the lift-to-drag ratio, by deflecting an angle δ_f. The fundamental cause lies in that δ_f leads to a series of changes of AoA of each part. Meanwhile, the gliding performance, like the pitch angle θ, the gliding velocity V as well as the gliding angle γ, are all altered. Certainly, their relationships are obviously nonlinear. When $\alpha_f > 0$, the downward-force moment is generated, and when $\alpha_f < 0$, the upward one is generated. As a result, θ in the condition of $\alpha_f > 0$ is smaller than that in $\alpha_f < 0$. Therefore, we can adjust δ_f to make $\alpha_f > 0$ to obtain a small θ or to make $\alpha_f < 0$ to obtain a large one. Besides, when $|\delta_f|$ is beyond some values (i.e., $30°$), $|\gamma|$ will increase with $|\delta_f|$, but the gliding velocity, as well as the horizontal gliding velocity, will decrease with the increase of $|\delta_f|$. These analysis results can help us design and control the robotic dolphin, for example, refining the fins with an excellent lift-to-drag ratio and predetermining a gliding angle for rapid gliding.

In the gliding process, pursing an expected pitch angle is very necessary, for example, for a high gliding efficiency using some suitable gliding angles. In this study, a closed-loop control method is proposed. Although slight undulation finally appears in the experiment, these results validate the fluke's capability for pitch regulation during gliding. Certainly, this is just a little part of the fluke. Combining a pair of flippers, the fluke can more strongly regulate the gliding performance. Distinct from the fixed wings of traditional underwater gliders, these controllable fins bring a lot of benefits to the gliding robotic dolphin. For instance, with the help of these controllable fins, the robot can perform several kinds of locomotion, such as dolphin-like swimming, gliding, turning in flippers, and so on, which make it more flexible in complex underwater environment. However, how to coordinate these locomotion patterns in 3-D space is another question worthy of further investigation. Moreover, constraint problems are ubiquitous in control systems, which need particularly be considered. Some research has been done on the constrained control of AUVs and obtained satisfied results [212], [213]. For example, Peng *et al.* presented an effective method for path following control of under-actuated AUV with velocity and input constraints [214]. As for the gliding robotic dolphin, it will also suffer from some constraints in an underwater environment, such as propulsive velocity, turning radius, pitching angle, and so on. In order to be safe in underwater environment, it is worth exploring how to construct an effective controller for the gliding robotic dolphin under some constraints.

5.6 CONCLUDING REMARKS

In this chapter, combined with the experimental data, the regulating effects of flippers and flukes on gliding motion are analyzed and discussed. On this basis, a heading controller based on differential flippers is proposed, which mainly includes three parts: (1) Considering that the gliding speed is necessary state feedback but difficult to measure in practice, a sliding mode observation is designed. (2) The control law is derived based on the backstepping method. (3) A PSO-based solver is designed for the nonlinear hydrodynamic characteristics of the flippers, which can be applied for the heading control while eliminating coupled but unnecessary roll and sideslip motion. Using the same system framework, the pitch-control method based on flukes is given. Finally, the effectiveness of the proposed attitude control methods is verified by simulations and experiments.

REFERENCES

[200] L. J. V. Uffelen, E. H. Roth, B. M. Howe, E. M. Oleson, and Y. Barkley, "A seaglider-integrated digital monitor for bioacoustic sensing," *IEEE J. Ocean Eng.*, vol. 42, no. 4, pp. 800–807, 2017.

[201] "XRAY Liberdade platform" [Online]. Available: http://auvac.org/platforms/view/241 (Accessed: 26-Nov-2016).

[202] Y. Shi, C. Shen, H. Fang, and H. Li, "Advanced control in marine mechatronic systems: A survey," *IEEE/ASME Trans. Mechatronics*, vol. 22, no. 3, pp. 1121–1131, 2017.

[203] J. Isern-Gonzalez, D. Hernandez-Sosa, E. Fernandez-Perdomo, J. Cabrera-Gamez, A. C. Dominguez-Brito, and V. Prieto-Maranon, "Path planning for underwater gliders using iterative optimization," in *Proc. IEEE Int. Conf. Robot. Autom.*, Shanghai, China, May 2011, pp. 1538–1543.

[204] A. Chalanga, S. Kamal, L. M. Fridman, et al., "Implementation of super-twisting control: Super-twisting and higher order sliding-mode observer-based approaches," *IEEE Trans. Ind. Electron.*, vol. 63, no. 6, pp. 3677–3685, 2016.

[205] P. Shi, M. Liu, and L. Zhang, "Fault-tolerant sliding-mode-observer synthesis of Markovian jump systems using quantized measurements," *IEEE Trans. Ind. Electron.*, vol. 62, no. 9, pp. 5910–5918, 2015.

[206] J. Yao, Z. Jiao, and D. Ma, "Extended-state-observer-based output feedback nonlinear robust control of hydraulic systems with backstepping," *IEEE Trans. Ind. Electron.*, vol. 61, no. 11, pp. 6285–6293, 2014.

[207] T. I. Fossen and J. P. Strand., "Tutorial on nonlinear backstepping: Applications to ship control," *Model. Identif. Control.*, vol. 20, no. 2, pp. 83–135, 1999.

[208] B. Sun, D. Zhu, and S. X. Yang, "A bioinspired filtered backstepping tracking control of 7000-m manned submarine vehicle," *IEEE Trans. Ind. Electron.*, vol. 61, no. 7, pp. 3682–3693, 2014.

[209] T. A. Johansen and T. I. Fossen, "Control allocation–A survey," *Automatica*, vol. 49, no. 5, pp. 1087–1103, 2013.

[210] F. Chen, R. Jiang, K. Zhang, B. Jiang, and G. Tao, "Robust backstepping sliding-mode control and observer-based fault estimation for a quadrotor UAV," *IEEE Trans. Ind. Electron.*, vol. 63, no. 8, pp. 5044–5056, 2016.

[211] J. Kennedy, "Particle swarm optimization," *Encyclopedia of Machine Learning*. Springer, pp. 760–766, 2011.

[212] H. Li and W. Yan, "Model predictive stabilization of constrained underactuated autonomous underwater vehicles with guaranteed feasibility and stability," *IEEE/ASME Trans. Mechatronics*, vol. 22, no. 3, pp. 1185–1194, 2017.

[213] Z. Peng, J. Wang, and J. Wang, "Constrained control of autonomous underwater vehicles based on command optimization and disturbance estimation," *IEEE Trans. Ind. Electron.*, vol. 66, no. 5, pp. 3627–3635, 2019.

[214] Z. Peng, J. Wang, and Q. Han, "Path-following control of autonomous underwater vehicles subject to velocity and input constraints via neurodynamic optimization," *IEEE Trans. Ind. Electron.*, vol. 66, no. 11, pp. 8724–8732, 2018.

CHAPTER 6

Gliding Motion Optimization for a Bionic Gliding Underwater Robot

6.1 INTRODUCTION

Gliding motion was realized by underwater gliders a long time ago and has been developed and successfully applied in ocean observatories and exploration. Several control strategies have been proposed for sawtooth gliding motion [10]. Tchilian *et al.* linearized the gliding dynamic model and applied a linear quadratic regulator (LQR) for pitch control to track the desired path [215]. Generally, the sawtooth gliding motion is only controlled by the internal movable mass and/or the net buoyancy. The movable mass directly adjusts the pitch angle, and then the gliding angle is indirectly adjusted. Different from traditional underwater gliders, the fins of bionic gliding underwater robots can serve as external control surfaces, further participating in sawtooth gliding motion control. However, current control approaches by external surfaces only concern either the pitch angle or the gliding angle. Another angle is passively controlled as an appurtenance. However, with the consideration of future underwater robots for more complex missions, the separate control of the pitch angle and gliding angle is indispensable, especially for some tracking or

DOI: 10.1201/9781003347439-6

docking applications in which onboard visual detection is necessary. Such complex gliding motion is rarely achieved by existing gliding underwater robots and gliding control strategies. The internal control surfaces can directly regulate the pitch attitude, while the external surfaces are capable of adjusting the external hydrodynamic forces. Thus, the movable mass and pectoral fins are suitable to serve as separate control surfaces. Overall, there are several challenges to the separate control of bionic gliding robots. First, the gliding motion of the gliding robot depends on the hydrodynamic forces. In general, a steady pitch attitude can lead to a certain steady velocity direction in the sawtooth motion. How to decompose the pitch attitude and gliding direction and how to coordinate the control surfaces for the complex motion are worth studying. Second, hydrodynamic forces are the key factors for gliding motion. In general, the hydrodynamic characteristics are mostly estimated by low-order polynomials pointing at the case of the smaller angle of attack (AOA). However, a large AOA will always exist when the desired gliding angle is independent of the desired pitch attitude. Therefore, the hydrodynamic characteristics in the case of AOA varying on a large scale should be considered for gliding control. In summary, future gliding control strategies demand the separate control of pitch attitude and gliding velocity, which requires two different actuators on a bionic gliding robot.

To reduce the energy consumption of the gliding motion, current studies can be summarized into three categories: motion parameter optimization, shape optimization, and optimal path planning. However, these studies aim to optimize gliding motion for underwater gliders with fixed wings. Most of them only considered the steady gliding motion without the dynamic part of gliding motion. The dynamic part is denoted as the transient gliding motion in this chapter. On one hand, transient gliding motion exists in each round of sawtooth gliding motion. It is necessary to optimize the transient gliding motion. On the other hand, the bionic gliding robot has a pair of active pectoral fins, which can serve as the extra control surface to regulate the gliding motion. The optimization by the pectoral fins does not conflict with the existing method. Therefore, the optimization strategy for transient gliding motion by pectoral fins is essential for gliding robots. The transient gliding motion optimization is assumed to be a Markov decision problem, and its complex system states can lead to the curse of dimensionality with conventional dynamic programming methods [216]. By bridging deep learning and reinforcement learning, deep reinforcement learning (DRL) utilizes neural networks as function approximators to address the curse of dimensionality [217]. It has been applied successfully in robotic control problems, including trajectory tracking [218], navigation strategy [219], and so on. In this chapter, the DRL is used to find an optimal deflecting rule of pectoral fins for the maximum gliding range.

In addition, there are some studies on gliding underwater robots with external control surfaces. Arima *et al.* validated that an underwater glider with independently controllable main wings has an admirable motion capability [158]. However, the optimizing capability of the external surfaces is not considered for gliding efficiency. In addition, the pectoral fins for bionic gliding robots are of bionic shape. The relations between the size, hydrodynamics, and optimizing capability of the pectoral fin are important for type selection and worth studying.

Aiming at the previously mentioned problems, the main objective of this chapter consists of two parts. The first is to explore the optimizing capability of pectoral fins for gliding motion and propose a DRL-based optimization strategy to improve the gliding range. Another is to propose a control strategy aimed at the separate control of pitch and gliding angles for a bionic gliding underwater robot with both internal movable mass and external fins. In the first part, the optimizing capability of the pectoral fins for the transient gliding motion is analyzed. The concept of transient gliding motion is presented, and its significance is illustrated. Six types of pectoral fins with different chords and wingspans are designed. Their hydrodynamics and optimizing abilities for transient gliding motion are analyzed and compared based on computational fluid dynamics (CFD) simulations. Another part is the DRL-based optimization strategy to improve the gliding range. We present an adversarial model including two competing gliding underwater robots to calculate the reward function. In the adversarial model, the two robots keep the same diving depths but different actions to calculate the improvement of the forward gliding distance more accurately. A two-stage reward function is also designed to obtain a superior solution. The training method is based on the double-deep Q-network (DQN) method and simulation results validate the proposed optimization strategy. Extensive aquatic experiments are carried out to further validate the effectiveness of the proposed strategy. In the second part, the separate control strategy consisting of a pitch controller and an AOA controller is proposed, in which the former is designed by the backstepping method and the latter by the MPC method. Gliding motion tracking of a desired gliding angle with a steady pitch attitude achieved by a bionic gliding underwater robot is novel and contributes to the development of underwater visual missions for autonomous underwater vehicles. For this, the gliding dynamics are decomposed into pitch and velocity terms, based on which the two controllers are specifically designed for separate control. The movable mass and the pectoral fins are effectively coordinated to regulate the pitch angle and the AOA, respectively. Simulations, including separate control and path following with the desired pitch attitude, are conducted to verify the feasibility and superiority of the proposed control strategy. Furthermore, an underwater gliding measurement and control system is designed and implemented for experimental verifications. The gliding states are measured by processing the images captured by a binocular camera. Based on the real-time feedback, extensive aquatic experiments are carried out to perform the separate and concurrent control of pitch angle and gliding angle as well as the gliding optimization, which further validates the effectiveness of the proposed strategies. This chapter is dedicated to the motion optimization and complex motion control of gliding underwater robots and supplies clues to performing diversified underwater missions in addition to visual perception and autonomous docking by efficient gliding motion.

The remainder of this chapter is organized as follows. Section 6.2 introduces the prototype and dynamic modeling of the bionic gliding underwater robot as well as the distinction of the transient gliding motion. Section 6.3 analyzes the hydrodynamics and regulating capability of several pectoral fins for comparison. The DRL-based optimization strategy is detailed in Section 6.4. Section 6.5 details the model decomposition and the gliding control strategy. The numerical simulations encompassing gliding optimization and control are provided in Section 6.6. Section 6.7 presents the design of the gliding measurement and control system and experiments. Finally, Section 6.8 gives concluding remarks.

6.2 BIONIC GLIDING UNDERWATER ROBOTIC SYSTEM

6.2.1 Bionic Gliding Underwater Robot

In this section, we first introduce a bionic gliding underwater robot for subsequent verification of the gliding optimization and separate control. The bionic glider is designed and implemented by mimicking a whale shark inspired by its broad and flattened shape [220] as well as its capability for long-distance travel [221]. As illustrated in Figure 6.1, it is approximately

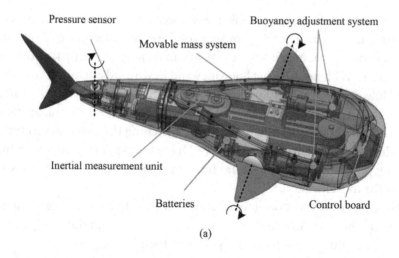

Pressure sensor

Buoyancy adjustment system

Movable mass system

Inertial measurement unit

Batteries

Control board

(a)

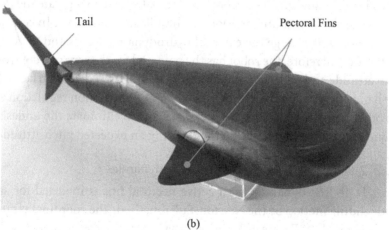

Tail

Pectoral Fins

(b)

FIGURE 6.1   A whale shark–like bionic underwater glider. (a) Mechanical struc-ture. (b) Prototype.

0.47 m long and 0.26 m wide and weighs approximately 2.2 kg. A pair of pectoral fins and a tail serve as the external control surfaces to regulate gliding motion or propel itself, similar to a real fish. The pectoral fins can pitch up or down in the range of ±90° actuated by two servos so that they can be deflected at the same or different angles as required. The shapes of the fins are designed based on the real pectoral fins of a whale shark to improve maneuverability. The internal actuators consist of a buoyancy adjustment system and a movable mass system. The buoyancy adjustment system is used to change the buoyancy for the driving force of gliding. The movable mass system can shift a movable mass longitudinally for attitude regulation. With negative net buoyancy and downward pitching, the glider can glide downward and forward driven by hydrodynamic forces. When arriving at a target depth, the robot changes its net buoyancy and attitude and then glides upward and forward. By switching the two states in turn, the glider achieves two-dimensional (2-D) sawtooth gliding motion. In the gliding process, deflected pectoral fins can change the hydrodynamics to affect the movement.

For this prototype, both movable mass and pectoral fins are capable of adjusting the pitch attitude for gliding motion. Although they are capable of attitude and gliding motion regulation, their driving mechanisms are different. The movable mass produces internal torque to regulate the pitch attitude so that the gliding motion can be adjusted indirectly. In contrast, the pectoral fins change the external hydrodynamic forces and the external torque. Therefore, the robot has the potential for the separate control of pitch attitude and gliding motion by utilizing the two control surfaces. On one hand, pectoral fins can be employed for gliding motion regulation. On the other hand, the movable mass can be used to eliminate the undesired torque derived from the fins, further holding an expected pitch attitude.

6.2.2 2-D Gliding Dynamics and Hydrodynamics

The 2-D gliding dynamics with active pectoral fins is modeled for sawtooth gliding motion by the Newton-Euler method. The coordinate frames are illustrated in Figure 6.2. The inertia frame is fixed and defined as $C_g = o_g x_g y_g$ with its origin o_g at the glider's initial position, axis $o_g y_g$ along the direction of gravity, and axis $o_g x_g$ along the direction of forward motion. The body frame is defined as $C_b = o_b x_b y_b$ with its origin o_b at the robot's center of buoyancy and axis $o_b x_b$ along the front of the robot. When the robot is in the initial position, the body frame and the inertia frame coincide. The pectoral fin frame is defined as $C_p = o_p x_p y_p$, with its origin o_p at the fins' center of gravity in body frame plane, and axis $o_p x_p$ along the front of the fins. The linear velocity of the robot in the inertia frame is defined as $v_g = (v_x, v_y)^T$. The linear velocity vector in the body frame is defined as $v_b = (v_1, v_2)^T$ and the angular velocity is ω. The pitch angle θ is the angle between $o_b x_b$ and $o_g x_g$. The kinematics for 2-D gliding motion are formalized by

$$v_x = v_1 \cos\theta + v_2 \sin\theta$$
$$v_y = -v_1 \sin\theta + v_2 \cos\theta \; . \tag{6.1}$$

The mass of the robot is divided into the movable mass m_m, the water sucked into the buoyancy adjustment system m_w, and the rest of mass m_b, whose location vectors with respect to C_b are $(l_{x,m}, l_{y,m})^T$, $(l_{x,w}, l_{y,w})^T$, and $(l_{x,b}, l_{y,b})^T$, respectively. The total inertia matrix is $\mathbf{M} = (m_b + m_m + m_w)\mathbf{I} + \mathbf{M}_f = diag\{m_1, m_2\}$, in which $\mathbf{M}_f$ denotes the additional inertia matrix from the surrounding water. The external forces include gravity, buoyancy, and hydrodynamic forces. The net buoyancy $m_0 g$ represents the difference in buoyancy and gravity. It is assumed that the center of gravity is on the axis $o_b y_b$ in the initial state.

Considering that the desired gliding motion may be gliding upward/downward and backward, the gliding angle θ_g and the AOA α_b are defined in the range from $-\pi$ to π as

$$\theta_g = \text{sgn}(v_y) \cdot \arccos\left(\frac{v_x}{\sqrt{v_x^2 + v_y^2}}\right) \tag{6.2}$$

$$\alpha_b = \text{sgn}(v_2) \cdot \arccos\left(\frac{v_1}{\sqrt{v_1^2 + v_2^2}}\right) . \tag{6.3}$$

By Newton's law, the dynamic model for 2-D gliding motion can be obtained as

$$\dot{v}_1 = m_1^{-1}(-m_2 v_2 w + m_0 g \sin\theta + r_x w^2 - r_y \dot{\omega} - D_b \cos\alpha_b$$
$$+ L_b \sin\alpha_b + f_{1,p})$$
$$\dot{v}_2 = m_2^{-1}(m_1 v_1 w - m_0 g \cos\theta + r_y w^2 + r_x \dot{\omega} - D_b \sin\alpha_b$$
$$- L_b \cos\alpha_b + f_{2,p})$$
$$\dot{\theta} = \omega$$
$$\dot{\omega} = J^{-1}((m_2 - m_1)v_1 v_2 + M_b - r_x v_1 \omega - r_y v_3 \omega + M_p - r_y \dot{v}_1$$
$$+ r_x \dot{v}_2 - r_y g \sin\theta - r_x g \cos\theta) \tag{6.4}$$

where J is the rotational inertia coefficient, D_b is the drag force, L_b is the lift force, M_b is the hydrodynamic moment, $\alpha_b = \arctan(v_2 / v_1)$ is the

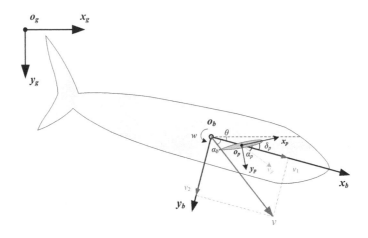

FIGURE 6.2   Coordinate frames of the bionic gliding underwater robot.

attack angle of the body, $f_{1,p}$ and $f_{2,p}$ are the component forces that pectoral fins exert on the body. r_x and r_y are calculated by

$$r_x = m_b l_{x,b} + m_m l_{x,m} + m_w l_{x,w} = m_m u_m + m_0 l_{x,w}$$
$$r_y = m_b l_{y,b} + m_m l_{y,m} + m_w l_{y,w} \qquad (6.5)$$

where u_m is the displacement of the movable mass.

Based on the quasi-steady model [222], the hydrodynamic forces of the body are acquired by

$$M_b = \rho \|v_b\|^2 H_M(\alpha_b) + k_q \omega$$
$$D_b = \rho \|v_b\|^2 H_D(\alpha_b) \qquad (6.6)$$
$$L_b = \rho \|v_b\|^2 H_L(\alpha_b)$$

where the hydrodynamic coefficients H_M, H_D, and H_L, and the rotation damping coefficient k_q can be estimated by CFD simulations.

The velocity vector of the pectoral fins with respect to C_p is defined as $v_p = (v_{p1}, v_{p2})^T$. The deflected angle of the pectoral fins is defined as δ_p. The distance between o_b and o_p is $l_{x,p}$ so that the velocity of the pectoral fins can be calculated as

$$v_{p1} = v_1 \cos\delta_p + (v_2 - w \cdot l_{x,p})\sin\delta_p$$
$$v_{p2} = -v_1 \sin\delta_p + (v_2 - w \cdot l_{x,p})\cos\delta_p \qquad (6.7)$$

With the attack angle $\alpha_p = \arctan(v_{p2}/v_{p1})$, the forces of the fins are calculated as

$$D_p = \rho \|v_p\|^2 H_{Dp}(\alpha_p)$$
$$L_p = \rho \|v_p\|^2 H_{Lp}(\alpha_p)$$
$$f_{1,p} = -D_p \cos(\alpha_p - \delta_p) + L_p \sin(\alpha_p - \delta_p) \cdot \qquad (6.8)$$
$$f_{2,p} = -D_p \sin(\alpha_p - \delta_p) - L_p \cos(\alpha_p - \delta_p)$$
$$M_p = l_{x,p} \cdot f_{2,p}$$

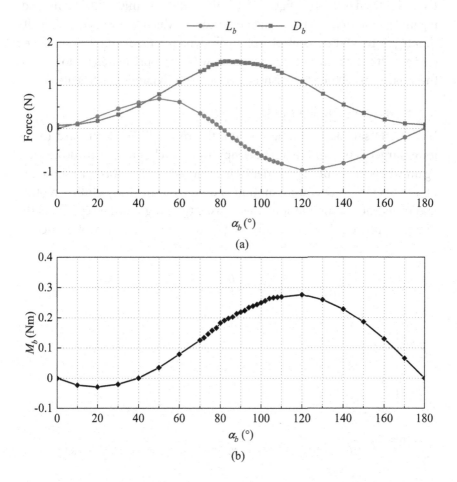

FIGURE 6.3 Hydrodynamic data of the body part by CFD simulations. (a) Hydrodynamic forces. (b) Hydrodynamic torque.

The CFD method is employed for the evaluation of the hydrodynamic coefficients, which is a common method for gliding motion [223]–[224]. For the robotic body, the hydrodynamic data are simulated when α_b varies from 0° to 180°. For the pectoral fins, the cases of α_b varying from 0° to 90° are simulated with the consideration of the expected range of 90° during control.

All CFD simulations are performed by the Flow Simulation software in SolidWorks [225]. In the simulation, an unstructured mesh is formed. The velocity inlet with $v = 0.3$ m/s and the pressure outlet are both double the body length away from the robot. The fluid is assumed to be incompressible and steady, and the $k - \omega$ shear–stress–transport turbulence model with low-Re corrections is adopted. The simulated hydrodynamic data are shown in Figures 6.3 and 6.4. As the figures show, the hydrodynamics of the bionic gliding underwater robot have strong nonlinearity, which increases the difficulty of integrating gliding control. For hydrodynamic modeling, the simulated data are fitted by high-order polynomials with appropriate weights to serve as the evaluations of the previously described hydrodynamic coefficients.

6.2.3 Transient Gliding Motion

In the 2-D sawtooth gliding motion, the robot glides downward to the lower target depth and then glides upward, switches to gliding downward again when the upper target depth is achieved, and so forth, as shown in Figure 6.5. The 2-D gliding motion is divided into steady gliding motion and transient gliding motion. The transient gliding motion represents the dynamic process from a steady motion to another, which often occurs

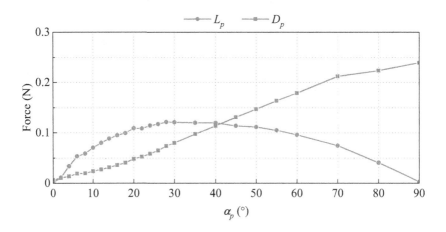

FIGURE 6.4 Hydrodynamic data of a single pectoral fin by CFD simulations.

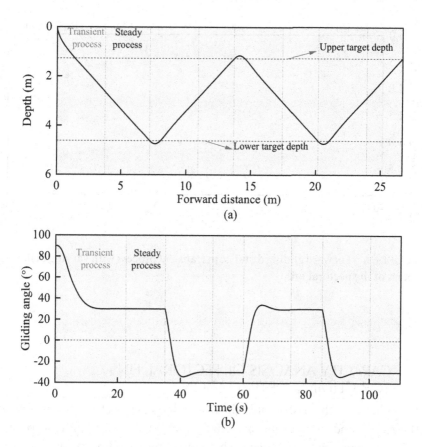

FIGURE 6.5 Steady and transient motion in the 2-D sawtooth gliding motion. (a) Gliding path. (b) Gliding angle.

when the gliding direction changes. In the transient process, especially when starting from the initial state (the first transient process), the gliding angle gradually changes from 90° to a steady-state value, in which the gliding path is steeper, which causes the loss of forward distance. Taking the downward gliding process as an example, we calculate the forward distances under several diving depths in the case of different δ_p. As seen in Figure 6.6, the pectoral fins can also evidently change the forward distance. It is worth noting that the optimal deflected angles corresponding to the maximal forward distances are different with diving depth, which means that an optimization deflecting rule of the active pectoral fins probably exists to maximize the forward gliding distance. This chapter aims to explore the optimization deflecting rule.

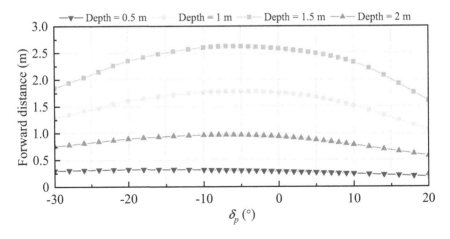

FIGURE 6.6   Forward gliding distance in transient process with varying deflected angles of the pectoral fins.

6.3 CAPACITY ANALYSIS OF PECTORAL FINS FOR GLIDING OPTIMIZATION

Different from the underwater glider with large wings, the pectoral fins of the gliding underwater robot are smaller and have bionic shapes. Therefore, it is indispensable to investigate what kind of pectoral fins can achieve better optimization for gliding efficiency. Several fins with different chords and wingspans are designed and analyzed for gliding optimization on the basis of CFD simulations.

6.3.1 Pectoral Fins Design

Considering the dimension of the gliding underwater robot, we design some typical fins, as shown in Figure 6.7. P_1 to P_6 represent the six fins with different chords and wingspans, respectively. The NACA 0012-64 airfoil is adopted. Their shapes and surfaces are generated with similar parameters in SolidWorks. They involve the cases of the same chords, the same wingspans, the same aspect ratios, and similar section areas as tabulated in Table 6.1.

6.3.2 Hydrodynamics of the Fins

The forces that the fins exert on the robot body are mainly derived from the hydrodynamic forces of the fins. To explore the optimization capabilities,

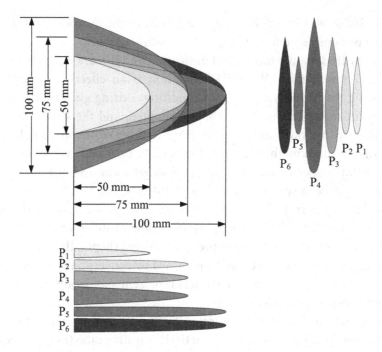

FIGURE 6.7 Three views drawing of the six designed pectoral fins with different chords and wingspans.

TABLE 6.1 Detailed Sizes of the Six Designed Pectoral Fins

Number	Chord	Wingspan	Aspect ratio	Section area
P_1	50 mm	50 mm	1	1667 mm²
P_2	50 mm	75 mm	1.5	2500 mm²
P_3	75 mm	75 mm	1	3750 mm²
P_4	100 mm	75 mm	0.75	5000 mm²
P_5	50 mm	100 mm	2	3333 mm²
P_6	75 mm	100 mm	1.33	5000 mm²

their hydrodynamics are first analyzed. Generally, the hydrodynamic forces are expressed by first-order or second-order terms of the attack angle, which is available only when the attack angle varies in a small scope. For the transient gliding process, the attack angle of the fins may vary from $-90°$ to $90°$. Therefore, complete models of the hydrodynamic forces should be considered for gliding motion optimization.

The hydrodynamic forces, including drag and lift, are estimated based on CFD simulations. The CFD simulation is an effective approach to acquiring the hydrodynamic force conditions during gliding. The quasi-steady lift-resistance model method is adopted, and the simulations are executed with Flow Simulation software in SolidWorks. In simulations, the unstructured mesh is formed, the velocity inlet with $v = 0.3$ m/s and the pressure outlet are both five times the length away from the fin, and the fluid is supposed to be incompressible and steady. Since the fins are usually turned slowly during gliding, the dynamic effect is not considered in this chapter. The lift-and-drag forces are estimated at different attack angles in the range of $0°$ to $90°$. Due to the symmetrical shapes, the forces, when the attack angle is negative, can be obtained by conversion. The results are shown in Figure 6.8, where the scatters represent simulation data and the lines represent fitted curves.

As shown in Figure 6.8, indicated by P_1, P_2, and P_5 with the same chords, the forces become greater, but the lift-to-drag ratio (LDR) decreases as the wingspan increases. For P_2, P_3, and P_4, which have the same wing-spans, the forces and LDR both increase as the chord increases. For P_1 and P_3 with the same aspect ratio, the magnitude of the forces is proportional to the cross area. In the case of similar cross areas such as P_3 and P_5 or P_4 and P_6, their magnitudes of the forces are similar, but the one with a larger aspect ratio has larger hydrodynamic forces when the attack angle is close to zero. In summary, the magnitude of the forces is proportional to the cross area. When the cross area is constant, the greater the aspect ratio is, the faster the hydrodynamic forces increase along with the increased attack angle. The LDR is inversely proportional to the aspect ratio when the cross area is constant and is proportional to the cross area when the aspect ratio is constant.

6.3.3 Optimizing Capability Analysis

Based on the hydrodynamics of the six fins, their capabilities for transient gliding optimization are analyzed. The forward distances with diving depths of 0.5 m, 1 m, 1.5 m, and 2 m from the initial equilibrium state are calculated in the cases of different δ_p. The results are shown in Figure 6.9. The curves of forward distances in the transient process show great similarity with the lift curves, which indicates that the performance of the transient process is

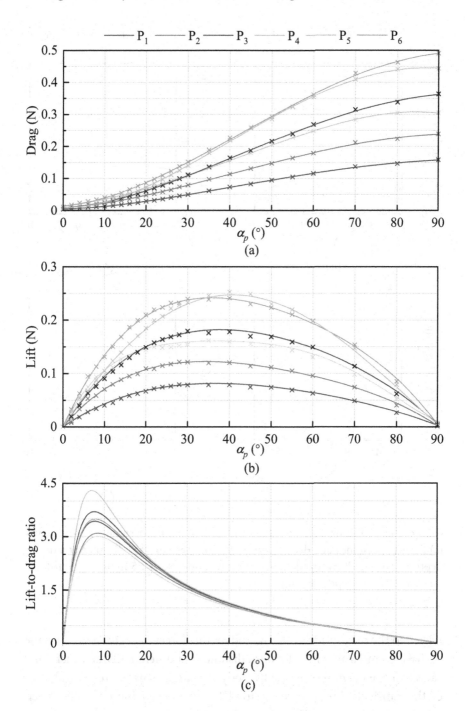

FIGURE 6.8 Hydrodynamics of the designed pectoral fins. (a) Drag force curves. (b) Lift force curves. (c) LDR curves.

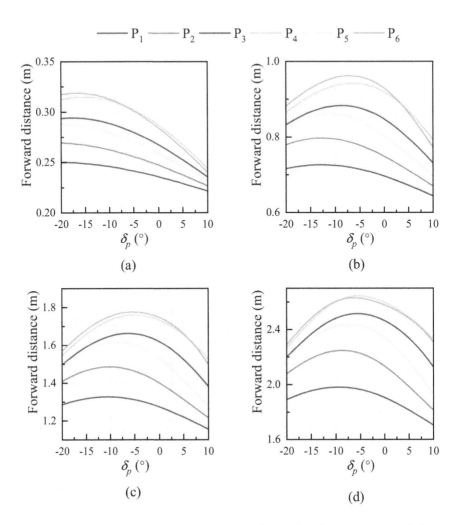

FIGURE 6.9 Gliding forward distance results of the designed pectoral fins. (a) Forward distance at a diving depth of 0.5 m. (b) Forward distance at a diving depth of 1 m. (c) Forward distance at a diving depth of 1.5 m. (d) Forward distance at a diving depth of 2 m.

related to the magnitude of the hydrodynamic forces of pectoral fins. P_4 and P_6 have a similar maximal forward distance but different LDRs, which indicates that the magnitude of the hydrodynamic forces has a greater influence on the transient gliding process than LDR. Therefore, to improve gliding performance, pectoral fins with large hydrodynamic forces are recommended.

In the transient gliding process, the deflected angle that maximizes the forward distance when diving to 0.5 m cannot increase the distance when diving 1 m; in other words, the deflected angles corresponding to the maximal forward distance at each diving depth are different. This reveals that the deflected angle must be active to maximize the forward distance in the whole transient process.

6.4 DRL-BASED GLIDING OPTIMIZATION STRATEGY

Similar to underwater gliders, gliding efficiency depends on the times of switching, which relates to the achievable forward distance in once-through descent or ascent. The optimization of the forward distance in the transient process can be regarded as a dynamic planning problem, that is, choosing a δ_p at each time step to maximize the total forward distance at a certain diving depth. However, the dynamics exist in four numerically continuous states, and the pectoral fins can deflect in a wide range, which causes the "curse of dimensionality" problem by normal dynamic planning methods. Distinct from conventional optimization algorithms, the heuristic DQN not only is appropriate for numerically continuous gliding states but is also able to train a neural network serving as the controller, with no need for online solving. In this chapter, we develop an adversarial DQN-based DRL method to improve the forward gliding distance in the transient gliding process via active pectoral fins. First, discretization of the gliding dynamics is performed. Then, the reward function is designed by the difference between the states of two competing robots. The training method is designed based on the double DQN. The proposed optimization strategy framework is shown in Figure 6.10.

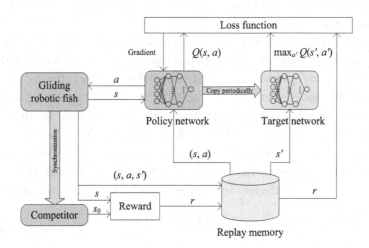

FIGURE 6.10   Framework of the proposed optimization strategy based on DQN.

6.4.1 Discretization

The optimization objective is the total forward gliding distance at a certain diving depth, that is, the sum of the forward distance values in each step. Therefore, the gliding dynamics is discretized with a variable sample time interval in which the diving depth of each step is constant, while a threshold value for the interval is also set to avoid an overlarge sample interval. The diving depth and forward distance of each step and the whole process are defined as Δ_y, Δ_x, s_y, and s_x, respectively. The time step is represented by k. Then, the sample time interval $t_s(k)$ is defined as

$$
t_s(k) = \begin{cases} t_{th} & , \quad u_{depth}/v_y(k) > t_{th} \\ u_{depth}/v_y(k) & , \quad \text{others} \end{cases} , \tag{6.9}
$$

where t_{th} is the threshold value and u_{depth} is the unit diving depth. The state is $s = (v_1(k), v_2(k), \theta(k), w(k))^T$, the state of the next step is $s' = (v_1(k+1), v_2(k+1), \theta(k+1), w(k+1))^T$, and the action is $a = \delta_p$. The nonconvex optimization problem is expressed as

$$
\max \sum_{k=1}^{n} \Delta x(k) , \tag{6.10}
$$

$$
\text{s.t.} \quad \delta_{min} \leq a \leq \delta_{max}
$$

where δ_{min} and δ_{max} are the threshold values of a and n is the number of total steps.

6.4.2 Reward Shaping

In each step, the forward distances corresponding to different actions are of too small difference to effectively train the network. Thus, we introduce a competing robot that has the same dynamics as the gliding underwater robot. The only difference is that the pectoral fins of the competing robot are not deflected. The two robots start from the same initial state, and in each step, the diving depth of the gliding underwater robot is assigned to the competing robot for synchronization. It should be noted that the states of the two robots are dissimilar with a high probability. For the competing robot, its state is defined as s_0, the state of the next step is s_0', the forward distances of each step and the whole process are Δ_{x0} and s_{x0}, respectively, and the sample interval is $t_{s0}(k) = \Delta_y(k)/v_{y0}(k)$. Then, we can acquire the forward distances in each step of the two robots as

$$
\Delta_x(k) = v_x(k)t_s(k) \tag{6.11}
$$

$$
\Delta_{x0}(k) = v_{x0}(k)t_{s0}(k) , \tag{6.12}
$$

where the forward velocities $v_x(k)$ and $v_{x0}(k)$ can be derived by Equation 6.2.

Two points need to be considered: (1) The forward distance in the current step is only related to the current state and has no connection with the current selected action. (2) $v_x(k)$ increases from a very small value so that the magnitude of $\Delta_x(k)$ cannot be regarded as consistent. Therefore, we divide the reward function into two parts. The first part is the improvement ratio of the forward distance in the next step as

$$r_1 = \mu_1 \frac{\Delta_x(k+1) - \Delta_{x0}(k+1)}{\Delta_{x0}(k+1)}, \tag{6.13}$$

where μ_1 is the normalization coefficient. The second part r_2 is related to the improvement ratio of forward distance in the whole process. To avoid repetition of the two parts, r_2 is only executed in the final step of each episode when the improvement ratio exceeds a threshold value λ_r.

$$r_2 = \begin{cases} \mu_2 , & \dfrac{s_x - s_{x0}}{s_{x0}} > \lambda_r , \\ 0 , & \text{others} \end{cases} \tag{6.14}$$

where μ_2 is an empirical coefficient. r_2 is used to further enhance the impact of episodes with a high improvement ratio.

6.4.3 Training Method

The optimal solution satisfies the Bellman equation. The reward of step k is $r(k)$ and the accumulated reward from step k to the end is $V_k = r(k) + r(k+1) + r(k+2) + \dots$. Then, the Bellman equation is expressed as

$$V_k = r(k) + \gamma V_{k+1}, \tag{6.15}$$

where γ is a discount factor. DQN utilizes neural networks to serve as the action-value function and estimate the Q value with experience. There are two neural networks in DQN, that is, the policy network and the target network. The policy network keeps evolving. The target network stays steady and updates its parameter by copying the policy network periodically. The policy network is defined as $Q(s, a; \phi)$ and the target network is defined as $Q(s, a; \phi^-)$, where the inputs are state and action, while the output is the Q value of the state–action input. The proposed training algorithm is presented in Algorithm 6.1.

From the beginning of each episode, the state of the gliding underwater robot s is initialized randomly in its state space and the state of the competing robot is assigned as $s_0 = s$. Then, the action is chosen by the ε-greedy algorithm for only the gliding underwater robot. The sample time of the competing robot $t_{s0}(k)$ is calculated by the diving depth of the gliding underwater robot in the current step. Thus, in each step, the two robots descend or ascend the same depth, and the difference in their forward distance is used for reward calculation. The episode is finished when the total diving depth S_y exceeds the set value d_{th}. In particular, at the ending step in each episode, r_2 will be calculated and added to the reward. We employ the following formula to calculate the target value as the double DQN [226] to avoid overestimation.

$$Y_t = r + \gamma Q(s', \arg\max_a Q(s', a; \phi), \phi^-),$$ (6.16)

where Y_t gives the target value of the double DQN at each step. Then, the policy network Q is updated by backpropagation with a loss function. The target network is assigned the same as the policy network every C steps. The preceding procedures are repeated until all episodes end.

ALGORITHM 6.1   Double DQN–based gliding optimization strategy by pectoral fins.

1: Initialize replay memory R to capacity N
2: Initialize policy network Q with random wights ϕ and target network $\hat{Q}$ with wights
   $\phi^- = \phi$
3: **For** episode $= 1, 2, ..., E$, **do**
4: Initialize state s randomly and competing state $s_0 = s$
5: **For** $k = 1, 2, ...,$ **do**
6: Select a random action a' with probability ϵ, otherwise, select
   $a' = \arg\min_a Q(s, a; \phi)$

7: Calculate s', Δ_y, t_{s0}, s_0', and r_1 in turn
8: $r = r_1 + r_2$ if $S_y > d_{th}$ else $r = r_1$
9: Store transition (s, a, r, s') in R
10: Sample random minibatch of transitions from R
11: $Y_t = r + \gamma Q(s', \arg\max_a Q(s', a; \phi), \phi^-)$

12: Update Q toward Y_t
13: Every C step set $\hat{Q} = Q$
14: Set $s = s'$
15: **Break for** if $S_y > d_{th}$
16: **End for**

6.5 SEPARATE CONTROLLER DESIGN

In this section, the dynamic model is decomposed, and then a backstepping-MPC control strategy is designed for separate pitch and gliding control.

6.5.1 Dynamic Model Decomposition

For the separate control of pitch and gliding angles, the dynamic model is decomposed into pitch and velocity terms. The pitch term is derived as the following equations, in which the states are pitch angle θ and pitch angular speed ω, while the input is the displacement of the movable mass u_m^-.

$$\dot{\theta} = \omega$$
$$\dot{\omega} = f_0 + b u_m$$
$$f_0 = J^{-1}((m_2 - m_1)v_1 v_2 + M_b + M_p - r_y \dot{v}_1 + r_x \dot{v}_2 \qquad (6.17)$$
$$-r_x v_1 \omega - r_y v_2 \omega - r_y g \sin\theta - m_0 l_{x,w} g \cos\theta)$$
$$b = -J^{-1} m_m g \cos\theta$$

In the velocity term, the pitch angle and its angular speed are regarded as a constant and zero, respectively. Then, the term can be derived as

$$\dot{X} = \begin{bmatrix} \dot{x}_1 \\ \dot{x}_2 \end{bmatrix} = \begin{bmatrix} f_1(x_1, x_2, u_p) \\ f_2(x_1, x_2, u_p) \end{bmatrix} \qquad (6.18)$$

$$f_1(x_1, x_2, u_p) = m_1^{-1} \rho \sqrt{x_1^2 + x_2^2}$$
$$\left\{ -[H_{Db}(\alpha_b) + H_{Dp}(u_p)]x_1 + [H_{Lb}(\alpha_b) + H_{Lp}(u_p)]x_2 \right\}$$
$$+ m_1^{-1} g m_0 \sin\theta$$
$$f_2(x_1, x_2, u_p) = m_2^{-1} \rho \sqrt{x_1^2 + x_2^2}, \qquad (6.19)$$
$$\left\{ -[H_{Lb}(\alpha_b) + H_{Lp}(u_p)]x_1 - [H_{Db}(\alpha_b) + H_{Dp}(u_p)]x_2 \right\}$$
$$- m_2^{-1} g m_0 \cos\theta$$

where x_1, x_2, and u_p represent v_1, v_2, and α_p for convenience, respectively.

6.5.2 Control Strategy

For 2-D gliding motion, the gliding angle is the difference between the AOA and the pitch angle, that is, $\theta_g = \alpha_b - \theta$. Therefore, gliding angle control can be equivalent to the control of AOA. Based on the decomposed dynamic models, pitch control is achieved using the movable mass and AOA control using pectoral fins. The control target is to track the desired gliding angle θ_{gd} and the desired pitch angle θ_d, that is, to track θ_d and the desired AOA $\alpha_{bd} = \theta_{gd} + \theta$ by control inputs u_m and u_p. Thus, the response of the pitch control is crucial for the whole control. For pitch control, because the hydrodynamic forces of the pectoral fins are modeled, the backstepping methodology is employed to rapidly eliminate the influence on the pitch attitude of the deflected fins. For AOA control, the controller is specifically designed by the MPC method to calculate the optimal deflected angle of the fins for the highly nonlinear system. The control system frame is depicted in Figure 6.11.

Because the hydrodynamic forces of the pectoral fins are much less than those of the body, the pectoral fins may be disabled to regulate the gliding motion when the robot inclines greatly. With the horizontal attitude, the pectoral fins can achieve regulation for a wide-range gliding angle. Thus, the proposed control strategy is more appropriate for tracking the desired gliding angle while maintaining a steady pitch angle for such a bionic gliding underwater robot.

6.5.3 Backstepping Pitch Controller

Based on the hydrodynamic model, the influence of deflected pectoral fins on pitch angle is known. Therefore, the pitch controller is designed by the backstepping methodology to eliminate the disturbance derived from the pectoral fins and track the desired pitch angle. The design procedures are as follows.

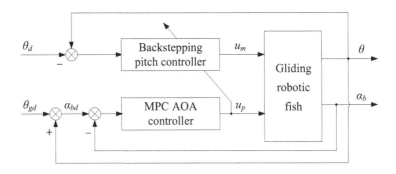

FIGURE 6.11   Backstepping-MPC gliding control strategy.

Step 1: The new state variables z_1 and z_2 are defined as

$$z_1 = \theta - \theta_d \tag{6.20}$$

$$\dot{z}_1 = \dot{\theta} - \dot{\theta}_d \tag{6.21}$$

$$z_2 = \dot{z}_1 + c_1 z_1, \tag{6.22}$$

where $c_1 > 0$ denotes a feedback gain and z_1 is the pitch angle error. Then, the first Lyapunov function V_1 is defined for z_1.

$$V_1 = \frac{1}{2} z_1^2 \tag{6.23}$$

$$\dot{V}_1 = z_1 \dot{z}_1 = z_1 z_2 - c_1 z_1^2 \tag{6.24}$$

It is obvious that $\dot{V}_1 \le 0$ if $z_2 = 0$.

Step 2: The second Lyapunov function is defined as

$$V_2 = V_1 + \frac{1}{2} z_2^2 \tag{6.25}$$

$$\dot{V}_2 = \dot{V}_1 + z_2 \dot{z}_2 = -c_1 z_1^2 + z_1 z_2 + z_2 (f_0 + b u_m + c_1 \dot{z}_1). \tag{6.26}$$

By making $z_1 z_2 + z_2 (f + b u_m + c_1 \dot{z}_1) = -c_2 z_2^2$, where $c_2 > 0$, the pitch control law is designed as follows:

$$
\begin{aligned}
u_m &= \frac{-f_0 - c_2 z_2 - z_1 - c_1 \dot{z}_1}{b} \\
&= \frac{-f_0 - (c_1 + c_2)\omega - (1 + c_1 c_2 + c_1^2)\theta + (1 + c_1 c_2)\theta_d}{b}
\end{aligned}
\tag{6.27}
$$

6.5.4 MPC-Based AOA Controller

The AOA controller is designed with the MPC method for a faster control response. First, the velocity model part is linearized at the current state point.

$$
\dot{\tilde{X}} = \begin{bmatrix} \dfrac{\partial f_1}{\partial x_1} & \dfrac{\partial f_1}{\partial x_2} \\[2ex] \dfrac{\partial f_2}{\partial x_1} & \dfrac{\partial f_2}{\partial x_2} \end{bmatrix}_{X=X_0} \tilde{X} + \begin{bmatrix} \dfrac{\partial f_1}{\partial u_p} \\[2ex] \dfrac{\partial f_2}{\partial u_p} \end{bmatrix}_{u_p = u_{p0}} \tilde{U} + \begin{bmatrix} f_{1,0} \\ f_{2,0} \end{bmatrix}, \tag{6.28}
$$

where X_0 and u_{p0} are the present state and the present control quantity, respectively $\tilde{X} = X - X_0$ and $\tilde{U} = u_p - u_{p0}$ represent the deviations of X

and u_p, respectively, and $f_{1,0}$ and $f_{2,0}$ are the values of $f_1(x_1, x_2, u_p)$ and $f_2(x_1, x_2, u_p)$ at the current time. By substituting the partial derivatives with simple symbols, the Equation 6.28 is written as

$$\dot{X} = \dot{\tilde{X}} = \begin{bmatrix} a_{11} & a_{12} \\ a_{21} & a_{22} \end{bmatrix} \tilde{X} + \begin{bmatrix} b_{11} \\ b_{21} \end{bmatrix} \tilde{U} + \begin{bmatrix} c_1 \\ c_2 \end{bmatrix}. \tag{6.29}$$

Then, the new state-space model is discretized with the sample time interval of t_s as

$$\tilde{X}(k+1) = \begin{bmatrix} a_{11}t_s + 1 & a_{12}t_s \\ a_{21}t_s & a_{22}t_s + 1 \end{bmatrix} \tilde{X}(k) + \begin{bmatrix} b_{11}t_s \\ b_{21}t_s \end{bmatrix} \tilde{U}(k) + \begin{bmatrix} c_1 t_s \\ c_2 t_s \end{bmatrix}, \tag{6.30}$$

where k is the current time step.

It is worth noting that it is strongly complex to derive the deviation of AOA by the deviation velocity. Thus, the discretized original state is derived as

$$X(k+1) = X(k) + \begin{bmatrix} a_{11}t_s & a_{12}t_s \\ a_{21}t_s & a_{22}t_s \end{bmatrix} \tilde{X}(k) + \begin{bmatrix} b_{11}t_s \\ b_{21}t_s \end{bmatrix} \tilde{U}(k) + \begin{bmatrix} c_1 t_s \\ c_2 t_s \end{bmatrix}, \tag{6.31}$$

which is written as

$$X(k+1) = X(k) + \mathbf{A}\tilde{X}(k) + \mathbf{B}\tilde{U}(k) + \mathbf{C}, \tag{6.32}$$

where $\mathbf{A}$, $\mathbf{B}$, and $\mathbf{C}$ represent the matrix at the corresponding position in Equation 6.31 for convenience.

With the preceding state model, the subsequent states of n_p steps are predicted with the controlling quantity $\{\tilde{U}(k), \tilde{U}(k+1), ..., \tilde{U}(k+n_p-1)\}$ as follows:

$$Y(k+1) = \hat{X}(k) + \Psi W + \Omega$$

$$Y(k+1) = \begin{bmatrix} X(k+1) \\ X(k+2) \\ ... \\ X(k+n_p) \end{bmatrix}, \hat{X}(k) = \begin{bmatrix} X(k) \\ X(k) \\ \vdots \\ X(k) \end{bmatrix}, W = \begin{bmatrix} \tilde{U}(k) \\ \tilde{U}(k+1) \\ ... \\ \tilde{U}(k+n_p-1) \end{bmatrix}, \tag{6.33}$$

$$\Psi = \begin{bmatrix} \mathbf{B} & \mathbf{O} & ... & \mathbf{O} \\ \mathbf{AB} & \mathbf{B} & ... & \mathbf{O} \\ ... & ... & ... & ... \\ \mathbf{A}^{n_p-1}\mathbf{B} & ... & ... & \mathbf{B} \end{bmatrix}, \Omega = \begin{bmatrix} \mathbf{C} \\ (\mathbf{A}+\mathbf{I})\mathbf{C} \\ ... \\ (\mathbf{A}^{n_p-1} + ...\mathbf{A}+\mathbf{I})\mathbf{C} \end{bmatrix}$$

The control target is to track the desired AOA. Thus, we define $k_a = \cot\alpha_{bd}$ so that $\min(\alpha_b - \alpha_{bd})^2$ is equivalent to $\min(x_1 - k_a x_2)^2$. Based on this, we further define

$$K = \begin{bmatrix} 1 & -k \\ -k & k^2 \end{bmatrix} \tag{6.34}$$

$$Q = diag\{\underset{n_p}{\mathbf{K},\mathbf{K},...,\mathbf{K}}\} . \tag{6.35}$$

Then, the optimal problem for AOA control is designed as follows:

$$J = \min_W \frac{1}{2}\left[Y^T QY + W^T RW\right], \tag{6.36}$$

where $\mathbf{R} \in \mathbb{R}^{n_p \times n_p}$ is a coefficient to reduce the chattering. The problem can be further written as

$$J = \min_W \left\{\frac{1}{2}W^T(\Psi^T Q\Psi + R)W + \left(\widehat{X} + \Omega\right)^T Q\Psi W\right\}, \tag{6.37}$$

which can be solved by the normal quadratic programming method. The control input $u_p(k)$ can be acquired by the optimal solution that represents the optimal deviation with the last-step input $u_p(k-1)$.

6.6 SIMULATION AND ANALYSIS

In this section, simulation results are offered to verify the effectiveness of the proposed optimization strategy and the separate controller. The influences of the parameters in the proposed strategy are also analyzed.

6.6.1 Training Results of the Gliding Optimization Strategy

Table 6.2 lists some parameters of the strategy in the training process. In practice, the overlarge deflected range of the pectoral fins may lead to instability. In addition, the resolution of the deflected angle is limited by the servo motors. Thus, in the learning process, the action space is defined as 0 to 160, corresponding to the deflected angles from −40° to 40° with the interval of 0.5°. The same fully connected networks with two hidden layers are adopted for the policy network and the target network. In the neural network, the input is the state s, the numbers of two hidden neurons are both 200, and the output is

TABLE 6.2 Feature Parameters of the Optimization Strategy

Parameters	Value	Parameters	Value	Parameters	Value
N	200000	μ_1	10	γ	0.95
u_{depth}	0.002 m	μ_2	500	minibatch	128
t_{th}	0.02 s	d_{th}	1 m	learning rate	0.00025
λ_r	0.02	C	50	ϵ	$e^{-0.0001k}$

the vector of Q values of all actions. In each episode, the gliding underwater robot starts from a random state in the small neighborhood of the equilibrium state and glides downward. The movable mass and net buoyancy are set as constants. When the diving depth exceeds 1 m, the episode ends.

Figure 6.12 shows the training results including the reward and the improvement ratio in each episode. The blue region represents the proposed algorithm, the orange region represents the algorithm without r_2, and the darker line represents the average value. The reward curves gradually converge as the episodes increase. For the results of the algorithm without r_2, the improvement ratio is lower although the reward is convergent. In contrast, the results of the proposed algorithm are also two-stage because of the two-stage reward function. In the first stage, the reward increases but is smaller. Then, the agent has learned something, and the improvement ratio reaches the target so that r_2 starts to perform and further increases both the reward value and improvement ratio. It suggests that the two-stage reward function is effective in finding the optimal solution. In addition, the curves of reward and improvement ratio are remarkably similar, which indicates that the designed reward function can precisely express the improvement of the gliding range. In particular, the curves oscillate at the end caused by the random initial states, which does not interfere with the convergence.

6.6.2 Dynamic Results of Gliding Optimization

The deflected rule of the pectoral fins from the proposed optimization strategy is employed in dynamic simulation. A rate limit is applied to avoid unnecessary vibrations in the rule. The steady deflected angle corresponding to optimal steady gliding motion is also simulated for comparison. With a net buoyancy of –0.03 kg and movable mass displacement of 0.04 m, the gliding underwater robot starts from the zero state and glides

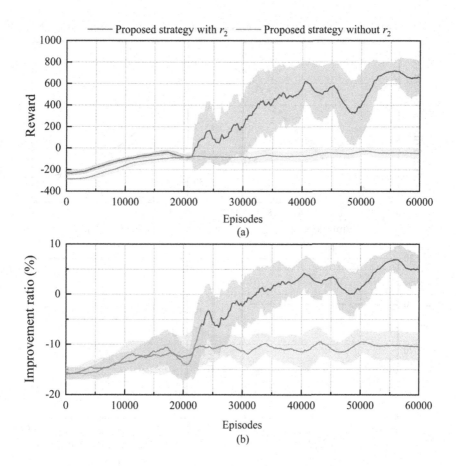

FIGURE 6.12   Training results of the proposed optimization strategy with and without r_2. (a) Reward curves. (b) Improvement ratio of the forward gliding distance in each episode.

downward. Three cases, in which pectoral fins do not deflect, deflect as the proposed optimization strategy, and deflect as the optimal steady value, are simulated. The results of the gliding path, gliding angle, improvement rate, and deflected rule are shown in Figure 6.13.

As seen from Figure 6.13, the deflected angle acquired by the proposed strategy is close to the threshold value first and increases gradually, which agrees with the former analysis of transient gliding motion. Compared with the cases of zero angle and constant angle corresponding to optimal steady motion, the proposed optimization strategy contributes to a smaller gliding angle and further forward gliding distance. From the improvement

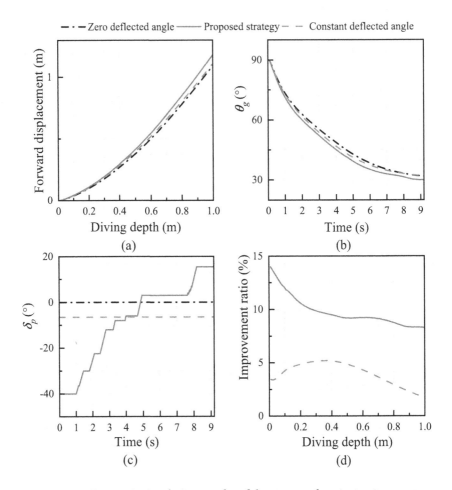

FIGURE 6.13 Dynamic simulation results of the proposed optimization strategy. (a) Gliding path. (b) Gliding angle. (c) Deflected angle of the pectoral fins. (d) Improvement ratio of forward gliding displacement.

ratio data of forward gliding distance at each diving depth, the proposed strategy can achieve an improvement ratio of more than 8.3%, which far outweighs the case of a constant deflected angle, which validates the optimization capability of the proposed strategy to improve the gliding range.

6.6.3 Separate Control

The parameters are derived from the actual robot, which are tabulated in Table 6.3. Considering the practical response speed of the servo motor, the rate limits are set to 0.02 m/s for u_m and 20°/s for δ_p.

TABLE 6.3   Feature and Control Parameters of the Bionic Gliding Underwater Robot

Parameters	Value	Parameters	Value	Parameters	Value
m_b	2 kg	$l_{y,b}$	0.01 m	n_p	10
m_m	0.2 kg	$l_{y,m}$	−0.06 m	c_1	2.3
$l_{x,w}$	0.1 m	$l_{y,w}$	0 m	c_2	10
$l_{x,p}$	0.04 m	k_q	−0.29	t_s	0.01
R	diag{0.00006, ...}				

The simulations consist of two parts. First, the performance of the proposed control strategy for pitch and gliding control is simulated and compared with the PID algorithm. Second, the performance for path following the horizontal attitude is validated.

In the first part, we verify the feasibility and property of the proposed gliding control strategy. Two PID controllers for pitch control and AOA control are designed for comparison, whose parameters are $(k_{p,pitch}, k_{i,pitch}, k_{d,pitch}) = (0.008, 0.0004, 4)$ and $(k_{p,aoa}, k_{i,aoa}, k_{d,aoa}) = (0.1, 0.00025, 0.8)$, respectively. The initial states of the robot are static with $\theta = 20°$ at the origin point in the inertia frame. then, the robot sucks water to descend while the controllers take effect. The desired gliding angle is 110°, and the desired pitch angle is 0° to simulate the situation in which the robot approaches a rear platform with a horizontal attitude. The simulation results are described in Figure 6.14.

As can be observed from Figure 6.14, with the proposed control strategy, both the pitch and gliding angles can reach the desired values rapidly. It should be noted that the results of the incremental PID algorithm are optimized with no overshoot and the shortest response time by our best efforts. Nonetheless, the response of the proposed control strategy is significantly faster than that of the PID algorithm.

6.6.4 Gliding Path Following

In this part, we utilize the proposed gliding strategy to follow a downward sine-shaped path from the origin point while always maintaining a horizontal attitude. The reference path is limited to gliding angles in the range of [70°, 110°]. The robot is first resting at the start point and then sucks water to follow the path with the proposed controller. The real-time reference gliding angle is calculated based on the gliding angle of the reference path at the current depth and the current position difference. The path-following results are described in Figures 6.15 and 6.16.

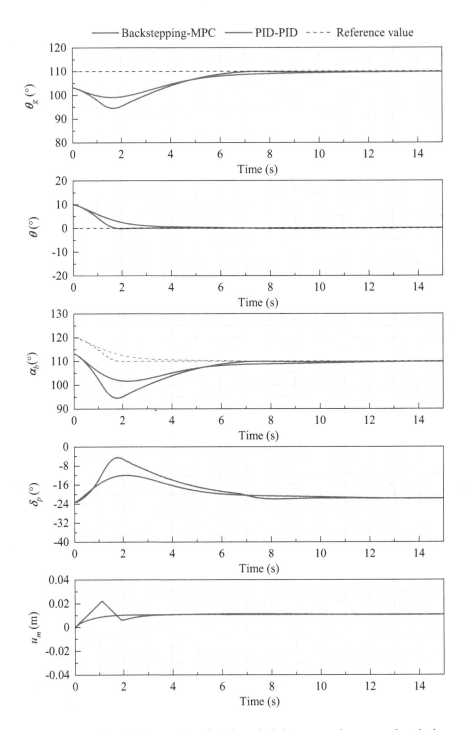

FIGURE 6.14   Simulation results of pitch and gliding control compared with the PID algorithm.

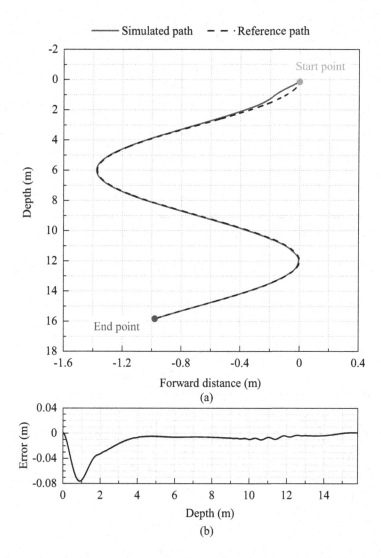

FIGURE 6.15 The trajectory and error result in path following. (a) Simulated path and reference path. (b) The following error of the horizontal distance.

Specifically, Figure 6.15 demonstrates that the proposed strategy is capable of path following and acquires remarkable performance. Because of the static initial state, the robot hardly precisely follows the path in the beginning. After descending approximately 4 m, the robot accurately reaches and follows the path with the error of horizontal distance less than 0.01 m. From Figure 6.16, it can be observed that the descent velocity is relatively stable, although without direct speed control. In addition, the pitch

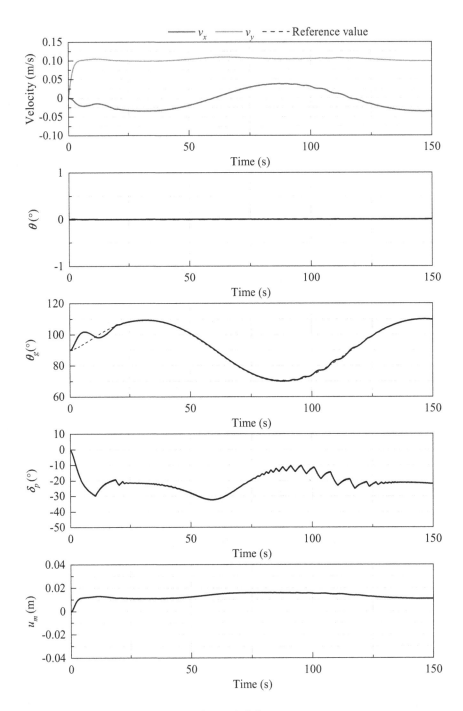

FIGURE 6.16 Simulation data in the path following.

angle is almost zero in the whole following process. The preceding results validate that the proposed controller is reliable and efficient for gliding path following with the horizontal attitude.

6.6.5 Discussion

As the results show, the proposed optimization strategy is feasible and effective for generating the deflecting rule of pectoral fins for efficient gliding. Through simulations, we find that some parameters observably affect the strategy. An excessively large u_r that may be unreachable will invalidate the two-stage reward function, while an excessively small u_r is incapable of obtaining the optimal solution. The large d_{th} causes a low speed of convergence, while the learning is inadequate for small d_{th}. This chapter validates the proposed strategy with a 1-m-deep gliding motion on account of limited computing power. The target depth of the whole gliding process and the depth achieved in the transient process should be taken into consideration for sufficient learning in practical applications. In the strategy, we acquire the optimal rule of pectoral fins with constant net buoyancy and constant displacement of movable mass, that is, the two variables are not the inputs of the network. The two variables can also adjust gliding motion so that if the variables change, the network should be retrained. Taking the two variables as the inputs is feasible but immensely increases the complexity of both network structure and training. In practical gliding travel, the net buoyancy and constant displacement of movable mass are set empirically, which indicates that a few networks trained by the proposed optimization strategy can serve as offline controllers to improve the gliding range of gliding underwater robots.

The proposed backstepping-MPC strategy responds faster than the PID–PID strategy, especially in pitch control. The probable reasons are analyzed. The controlled plant is a high-nonlinear multi-input, multi-output system. The active pectoral fins can directly affect the pitch angle, and the varying pitch angle also influences the gliding angle. The PID method performs control by the feedback, which does not utilize the influence that can be modeled. That the two PID controllers will affect each other may degrade the control performance. With regard to the system robustness for certain modeling uncertainties and external disturbances, some discussions are carried out. When the hydrodynamic parameters increase or decrease within 10%, the steady-state errors are no more than 1° in such simulation conditions. In addition, the control strategy is effective with a slight error when an external disturbance exists and is less than 10% of the hydrodynamic force of the pectoral fins.

6.7 EXPERIMENTAL VERIFICATION

Some experiments were performed to validate the practical capability of the proposed gliding optimization strategy and separate controller in an experimental pool with a depth of 1.3 m. All experiments were conducted in a designed underwater measurement and control system for gliding motion.

6.7.1 Measurement and Control System

It is difficult to measure the necessary gliding velocity in practical control, making the gliding optimization and control strategy difficult to verify. Although there have been some model-based observers for gliding velocity estimation, practical measurements are essential for the further verification of gliding control strategies. Therefore, we developed an underwater measurement and control system for gliding motion based on binocular vision as illustrated in Figure 6.17. The bionic gliding underwater robot pasted two rounded color markers on both sides of the buoyancy center. The binocular camera was installed on the sidewall of the pool, whose optical axis was perpendicular to the gliding plane. The upper computer calculated the real-time distances between the markers and the camera, by which the positions and velocities of the two markers were acquired. Then, the velocity of the buoyancy center, pitch angle, and pitch angular speed were calculated. In addition, the gliding underwater robot is equipped with a power sensor to measure the real-time consumed power. Based on the measured state values, the control quantities were derived by the proposed controllers, and corresponding control instructions were sent with radio frequency signals by the upper computer.

6.7.2 Sawtooth Gliding Experiments

This part aims to validate the proposed optimization strategy in the sawtooth gliding process. The experiments of descending and ascending gliding motions were separately performed because of the finite camera view. For the descending gliding process, the robot was first resting on the water surface with $m_0 = 0.03 \, \text{kg}$ and $u_m = -0.04 \, \text{m}$, simulating the state of gliding upward arriving at the surface. Then, the robot moved the mass and sucked water (make $m_0 = -0.03 \, \text{kg}$ and $u_m = 0.04 \, \text{m}$) to

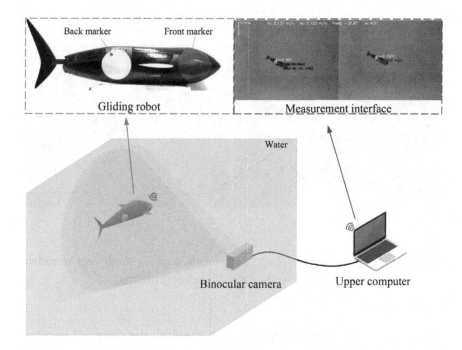

FIGURE 6.17   Gliding experiment scenes in the gliding measurement and control system.

dive. Meanwhile, the upper computer measured the gliding states and sent control instructions based on the trained network. The experiment ended when a 1-m diving depth was achieved. In addition, the case without optimization was also performed for comparison. For the ascending gliding process, the robot was first resting on the water bottom with $m_0 = -0.03$ kg and $u_m = 0.04$ m, simulating the state of gliding downward arriving at the bottom. Then, the robot moved the mass and emptied the water (make $m_0 = 0.03$ kg and $u_m = -0.04$ m). The frequency of recording the consumed power was 10 Hz. The frame rate of the camera was 16 Hz, which was also the control frequency. The gliding paths are illustrated in Figure 6.18. Meanwhile, Figure 6.19 shows the depth and power data in the descending process as a representative.

As indicated in Figure 6.18, the proposed strategy can improve the forward distance in the sawtooth gliding motion. In this trial, the forward distance increased by 0.15 m for descending motion and 0.12 m for ascending motion. In addition, as shown in Figure 6.19, the robot arrived at the

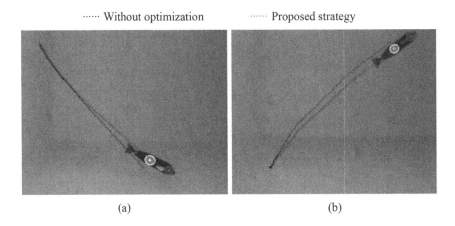

FIGURE 6.18 Sawtooth gliding path with and without optimization. (a) Descending gliding process. (b) Ascending gliding process.

target depth faster with optimization. Caused by the deflected pectoral fins, the consumed power is larger than that in the case without optimization. These results demonstrate that the proposed strategy is feasible and effective for improving the gliding range and speed in a round of sawtooth gliding motion.

6.7.3 Energy Consumption Statistics

The one-round gliding motion is used to represent the whole gliding process in this chapter, which consists of a descending gliding motion, an ascending gliding motion, and two state-transition processes involved. We tested the energy consumption of the one-round gliding motion in the cases with and without optimization. Each experiment was performed five times for statistical analysis. Thereinto, the descending motion and ascending motion were separately conducted as in the last subsection. The one-round energy consumption includes the consumption of the state transition on the water surface, the descending motion, the state transition on the water bottom, and the ascending motion. Based on the statistical results of the one-round gliding motion, the indexes of energy consumption and elapsed time are estimated, which denote the required average energy and time per meter for sufficiently long gliding travel, respectively. The statistical results and the performance estimation are tabulated in Table 6.4. Although the strategy increases the one-round energy

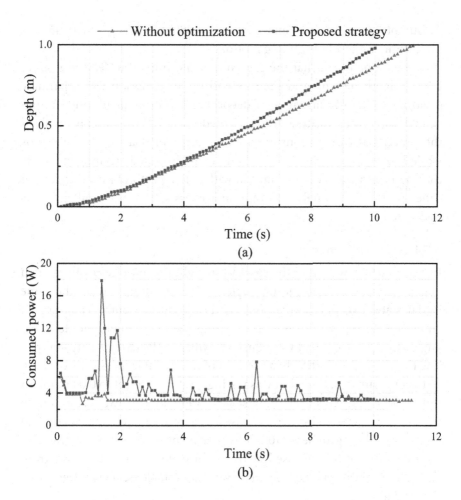

FIGURE 6.19 Depth and consumed power in the descending gliding process. (a) Depth. (b) Consumed power.

TABLE 6.4   Statistical Results and Estimation of Gliding Motion

	Without optimization	Proposed strategy
One-round energy consumption	114.4 J	124.0 J
One-round elapsed time	21.7 s	20.0 s
One-round forward distance	1.86 m	2.12 m
Energy-consumption index	61.5 J/m	58.25 J/m
Elapsed time index	11.67 s/m	9.4 s/m
Energy-saving rate	–	4.88%
Time-saving rate	–	19.45%

consumption, it can save approximately 4.88% of energy and 19.45% of travel time in the whole gliding motion.

It should be noted that the preceding experiments aimed for a target gliding depth of 1 m, and the sawtooth gliding motion was separately performed. The performance of deeper and continuous gliding motion should be further verified. The recorded energy consumption involved the functional energy consumed by the control board and the servos while holding their position. Although the proposed strategy increases the one-round energy consumption with active pectoral fins, the elapsed time and the functional consumption are less so that the total energy is optimized.

6.7.4 Separate Control Experiments

In experiments, we tested the cases of free gliding, only controlling pitch angle, only controlling gliding angle, and controlling both at the same initial state. Except for free gliding, the proposed controllers were all employed. In the case of only controlling the pitch angle, the deflected angle of the pectoral fins was constant, while in the case of only controlling the gliding angle, the displacement of the movable mass was constant. In the experiments, the robot was resting, and u_m was set to 0.04 m for an initial slope first. Then, the robot sucked water, and u_m was set to 0 m to perform experiments in the four cases. The desired pitch angle is 0°, and the desired gliding angle is 110°. The experimental results are illustrated in Figures 6.20 and 6.21. Caused by the robot inevitably moving away from the desired gliding plane, the trials with acceptable deviation were used for validation.

The initial pitch angle was approximately 15°. For free gliding (the red lines in Figure 6.21), the pitch angle and gliding angle nearly converged after several oscillations. The pitch angle could not converge to 0 as a result of the hydrodynamic torque of the bionic robot body so that the gliding angle was close to approximately 95°. For the cases of applying the proposed pitch controller (the cyan lines and the blue lines), the pitch angle curves have better smoothness and convergence. For the case of gliding angle control (the orange lines), even though the pitch angle was oscillating with a bias because of the pitch torque generated by the deflected pectoral fins, the gliding angle could also reach the desired value and converge. For the case of controlling both (the blue lines), the pitch angle and gliding angle could converge to the desired values. The obtained results

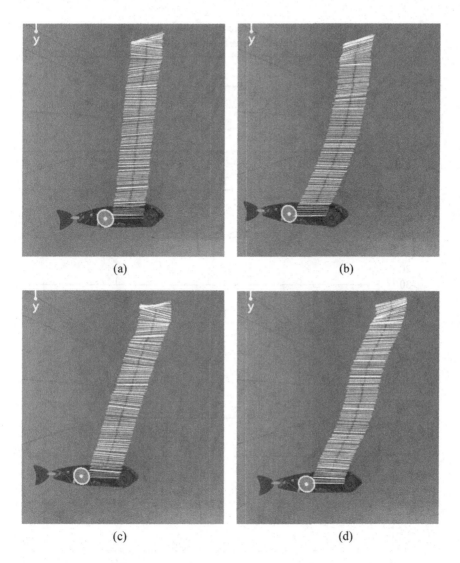

FIGURE 6.20 Experimental paths of four cases with white line segments whose directions represent pitch angles. (a) Free gliding. (b) Pitch control. (c) Gliding control. (d) Pitch and gliding control.

validate that the proposed controllers are of great capability not only for the separate control of pitch angle or gliding angle but also for concurrent control. Although measurement delay and error exist, the proposed strategy is still feasible and effective.

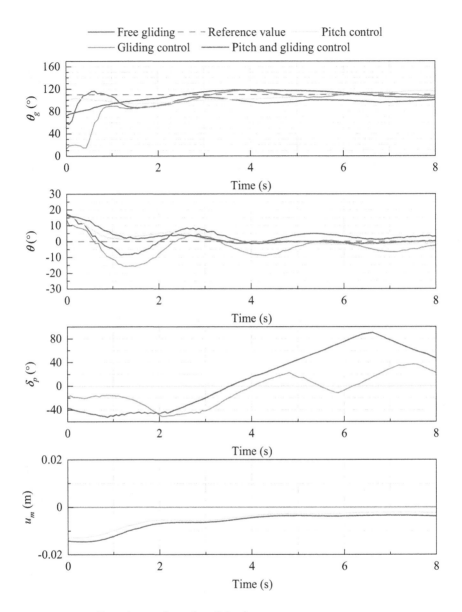

FIGURE 6.21 Experimental results of the four cases.

It is worth noting that the convergence speed in the experiments is lower than that in the simulations because delays in data processing and communication are inevitable and measurement error exists. In future applications, the measurement and control will certainly be implemented onboard via installed cameras or sensors in the robot, which can further improve the practical control performance. In addition, only the case of the desired horizontal attitude is performed in simulations and experiments. This is because a large hydrodynamic force in the horizontal direction can be generated when the robot glides with a nonzero pitch angle, which is just the mechanism of underwater gliding. However, the controllable hydrodynamic force generated by pectoral fins with small sizes is significantly less than that of the body, so gliding angle control may fail. If a gliding underwater robot has a relatively large-scale control surface, the proposed control strategy will be competent to track the desired gliding angle with a certain pitch angle.

6.8 CONCLUDING REMARKS

In this chapter, we proposed a DRL-based optimization strategy to improve gliding efficiency by active pectoral fins and an integrated gliding control strategy aimed at both pitch and gliding control for a bionic gliding underwater robot. First, the bionic gliding underwater robot is introduced, which is equipped with an internal movable mass and a pair of pectoral fins. A 2-D gliding dynamic model of the bionic gliding underwater robot is built. Its hydrodynamics in a large AOA range are simulated by the CFD method. Particularly, the concept of transient gliding motion is introduced. Second, several pectoral fins with different characteristics are designed to look for better hydrodynamic performance for gliding motion optimization. CFD results show that pectoral fins possessing greater hydrodynamic forces are recommended for gliding motion optimization. Third, an adversarial model including two competing gliding underwater robots is presented for the adequate calculation of the gliding range. A two-stage reward function is designed and then a double DQN–based optimization strategy is proposed to improve gliding efficiency with active pectoral fins. Fourth, based on the decomposed dynamic model, the control of gliding angle is converted to the control of AOA, and the pitch control law and AOA control law are designed with backstepping methodology and the MPC method, respectively. The pitch control is realized by the movable mass, while the AOA control is realized by the pectoral fins. Then, simulations are conducted to evaluate the convergence and effectiveness of the proposed strategies. For further verification, some experiments were conducted in the developed measurement and control system for underwater gliding motion. Finally, the experimental results indicate that the robot can save approximately 4.88% of energy and approximately 19.45% of travel time with the proposed optimization strategy. The capability of the proposed controller for separate and concurrent control of pitch angle and gliding angle was also confirmed. This study contributes to not only the design and optimization of active control surfaces but also the improvement of gliding efficiency and complex control for underwater vehicles.

REFERENCES

[215] R. D. S. Tchilian, E. Rafikova, S. A. Gafurov, and M. Rafikov, "Optimal control of an underwater glider vehicle," *Procedia Eng.*, vol. 176, pp. 732–740, 2017.

[216] G. Abhijit, "Reinforcement learning: A tutorial survey and recent advances," *INFORMS J. Comput.*, vol. 21, pp. 178–192, 2009.

[217] X. Wang, Y. Gu, Y. Cheng, A. Liu, and C. Chen, "Approximate policy-based accelerated deep reinforcement learning," *IEEE Trans. Neural Netw. Learn. Syst.*, vol. 31, no. 6, pp. 1820–1830, 2020.

[218] W. Zhang, J. Gai, Z. Zhang, L. Tang, Q. Liao, and Y. Ding, "Double-DQN based path smoothing and tracking control method for robotic vehicle navigation," *Comput. Electron. Agr.*, vol. 166, pp. 104985, 2019.

[219] H. Shi, L. Shi, M. Xu, and K. Hwang, "End-to-end navigation strategy with deep reinforcement learning for mobile robots," *IEEE Trans. Ind. Inform.*, vol. 16, no. 4, pp. 2393–2402, 2020.

[220] D. Rowat and K. S. Brooks, "A review of the biology, fisheries and conservation of the whale shark (*rhincodon typus*)," *J. Fish Biol.*, vol. 80, no. 5, pp. 1019–1056, 2012.

[221] H. M. Guzman, C. G. Gomez, A. Hearn, and S. A. Eckert, "Longest recorded trans-pacific migration of a whale shark (*rhincodon typus*)," *Mar. Biodivers. Rec.*, vol. 11, no. 1, pp. 1–6, 2018.

[222] M. R. Nabawy and W. J. Crowther, "On the quasi-steady aerodynamics of normal hovering flight part II: Model implementation and evaluation," *J. Royal Soc. Interface*, vol. 11, pp. 20131197, 2014.

[223] A. K. Lidtke, S. R. Turnock, and J. Downes, "Hydrodynamic design of underwater gliders using k-k_L-ω reynolds averaged Navier–Stokes transition model," *IEEE J. Ocean Eng.*, vol. 43, no. 2, pp. 356–368, 2017.

[224] Y. Singh, S. K. Bhattacharyya, and V. G. Idichandy, "CFD approach to modelling, hydrodynamic analysis and motion characteristics of a laboratory underwater glider with experimental results," *J. Ocean Eng. Sci.*, vol. 2, pp. 90–119, 2017.

[225] E. Bellos, D. Korres, C. Tzivanidis, and K. Antonopoulos, "Design, simulation and optimization of a compound parabolic collector," *Sustain. Energy Technol. Assess.*, vol. 16, pp. 53–63, 2016.

[226] H. Van Hasselt, A. Guez, and D. Silver, "Deep reinforcement learning with double Q-learning," in *Proc. Int. Conf. Artif. Intell. (AAAI)*, Washington, USA, 2016, pp. 2094–2100.

CHAPTER 7

Real-Time Path Planning and Following of a Gliding Underwater Robot within a Hierarchical Framework

7.1 INTRODUCTION

Efficient path planning has always been a challenging task for underwater robots and aimed at finding an optimal path from the initial to target points with full consideration of path length, safety, and smoothness. In recent years, many researchers have paid more attention to optimal path planning and environment adaptation. Pêtrès *et al.* [227] proposed a novel fast marching (FM)–based method to achieve path planning in view of wide continuous environments prone to currents. Aghababa [228] regarded path planning as a nonlinear optimal control problem and applied five intelligent evolutionary algorithms to compute the optimal path. Ataei *et al.* [229] generated offline paths for waypoint guidance of the autonomous underwater vehicle (AUV) and found a set of pareto-optimal solutions by considering four main criteria. Ma *et al.* [230] presented a novel ant colony system that incorporated an alarm pheromone trait to avoid many ineffective searches of large-scale problem space and verified

DOI: 10.1201/9781003347439-7

the proposed method via an established underwater environment model. Similarly, Han *et al.* [231] offered a complete-coverage path planning based on an ant colony algorithm for underwater gliders. Furthermore, with the popularity of artificial intelligence, some path planning studies based on learning algorithms have emerged. Martínez *et al.* [232] proposed a hierarchical Q-learning approach for motion planning of mobile robots, but the environment and state space were a bit simple, and the obstacles were static. Hu *et al.* [233] presented an incremental learning method to avoid obstacles using local-greedy policy-based Q-learning, and the adaptive kernel linear model was employed to store behavioral policies.

Moreover, there are many studies in terms of path following, tracking, and maneuvering which are used interchangeably in the literature. Fredriksen *et al.* [234] developed a way point maneuvering control based on a line-of-sight (LOS) law and a nonlinear control theory and gave the global exponential convergence properties. Li *et al.* [235] offered a two-time-scale path following control to simplify the implementation of practical engineering and considered the motion dynamics as fast and slow subsystems in singular perturbation control strategy. Liu *et al.* [236] presented a local desired trajectory to solve the problem of a large initial tracking error and used a single-input fuzzy model to determine the time interval. For environmental disturbances and systematical uncertainties, some researchers proposed many control methods from different points, such as the proportional integral (PI) sliding mode controller (SMC) [237], practical backstepping controller [238], and adaptive output-feedback control with prescribed performance [239].

In spite of many successful advances in path planning and following, there are still some limitations for practical engineering:

- *Path planning*: Most traditional path planning methods might only consider the issues in a static environment, which required exact map information before planning and usually could not achieve real-time planning. However, most real underwater environments are dynamic with movable obstacles, bringing great challenges to the planning algorithms. Consequently, it is significant to explore a utility algorithm to achieve real-time path planning in complex dynamic environments.

- *Path following and tracking*: Based on the previously surveyed literature, there were also many points that could be improved. First, some way points-based path following methods ignored the derivation of desired position, and the LOS law showed an unsmooth characteristic during the switching stage. Second, there existed the singularity of yaw angle in the surge law of some controllers based on the Lyapunov function. In practical underwater engineering, there will usually be a yaw angle error of $90°$, resulting in damage to the actuated mechanism with these methods.

For the gliding underwater robot, there are some extra challenges in terms of path planning and following. Due to the special motion modes, the kinematic and dynamic constraints of the gliding underwater robot need to be considered in path planning. With regard to the path following, the motions of the gliding underwater robot are relatively unique, which are different from traditional propeller-driven robots since their dynamic and energy models have been well studied.

In response to the earlier problems, we propose some methods to achieve the real-time path planning and following, which has not been addressed for a gliding underwater robot in previous literature, to the best of our knowledge. Compared with the existing work, the main contributions of this chapter conclude several aspects as follows:

1. A hierarchical engineering framework including path planning, improved LOS law, and path following is proposed, which provides theoretical support and technical foundation for many practical ocean tasks, such as ocean patrol, searching, or salvage.

2. A hierarchical deep Q-network (HDQN) is novelly proposed to achieve successful path planning, which not only satisfies the real-time requirement but also guarantees better performances in both static and unknown dynamic environments than traditional path planning methods.

3. A nonlinear adaptive backstepping controller (ABP) is particularly designed to follow the planned path, and the yaw singularity is technically avoided by applying barrier Lyapunov function (BLF). Meanwhile, the stability of closed-loop system is proved. From the obtained results, our designed controller also demonstrates excellent performances in following error and anti-interference ability.

The remainder of this chapter is organized as follows. Section 7.2 illustrates the overview of a gliding underwater robot. Section 7.3 elaborates the path-planning methodology of HDQN. Thereafter, the detailed derivation of controller design for path following is presented in Section 7.4. In Section 7.5, simulation and experimental results are offered and analyzed. Finally, Section 7.6 provides our conclusions.

7.2 OVERVIEW OF THE GLIDING UNDERWATER ROBOT

In this chapter, the gliding robotic dolphin is applied to verify the motion performance of the gliding underwater robot. Figure 7.1 illustrates the prototype of the gliding underwater robot. With regard to the dolphin-like motion, the flippers own the ability to produce the yaw moments through flapping a unilateral flipper or bilateral flippers differentially. Thus, multiple motions can be attained by combining with each other, and the robot can achieve some horizontal motion controls with designed multimodal, such as path following. Therefore, with the aid of flippers and caudal propulsive system, the robot can control the thrust and yaw moment with institutional decoupling, further laying the foundation for the planar path following.

7.3 PLANAR PATH PLANNING

7.3.1 Problem Statement and Network Architecture

The goal of planar planning is to generate a path from the initial to target points with obstacle avoidance and make the path as smooth and short as possible. First, by defining the positions of initial point, target point, and obstacles, we design a map with multiple obstacles. It should be noted that the obstacles' positions can be unknown or even dynamically movable. Besides, in order to facilitate the planning of obstacle avoidance paths, we expand the obstacle into a circle shape. These circles can overlap, so that various irregular shapes can be realized.

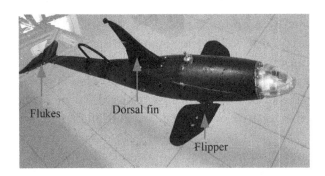

FIGURE 7.1 The prototype of the gliding underwater robot.

Considering the problem is an optimization with multiple objectives, we improve the deep reinforcement learning (DRL) methodology named HDQN to accomplish the planning task. In HDQN, we divide the planned path into two categories: target approach stage (TAS) and obstacle avoidance stage (OAS) from the point of different planning emphasis. For instance, we hope to pursue path smoothness more in TAS while focusing more on safety factor during OAS. It should be noted that only one network is used in the planning process, which is divided into TAS and OAS. More important, the two networks do not share parameter information, but their parameter update can be obtained in the same training episodes. This improvement can bring two benefits:

1. The size of the network can be designed much lighter than that using one single network, which not only makes it easier to converge to the global optimal solution due to fewer reward factors but also meets the requirement of real-time planning.

2. The comprehensive effect of the planned path in terms of smoothness, safety, path length and so on will be more satisfactory.

Furthermore, we consider path planning as a Markov decision process (MDP), which can be represented by states $s \in \mathcal{S}$, actions $a \in \mathcal{A}$, and transition $\mathcal{F}:(s,a) \rightarrow s'$. By choosing an action, the agent can arrive the next states, and the reward function $\mathcal{R}(s)$ can be recorded to quantify the current states. The goal is to maximize the cumulative reward in planning process further to obtain the optimal policy. The natural DQN adopts the fixed target network and experience replay module on the basis of traditional Q learning method, where the replay buffer can be denoted as $\mathcal{D}:(d_1,d_2,...,d_N),d_i=(s_i,a_i,r_i,s_i')$. By sampling the buffer to update the network, we can break the connection between data. Moreover, we denote $Q_i(s_i,a_i;\theta_i)$ as the approximate action-value function with θ_i, where i represents the TAS or OAS. The parameters θ_i will be updated by iteratively minimizing the loss function, which can be defined as

$$L_i\left(\theta_i\right)=\mathbb{E}\left(\left(y_i-Q_i\left(s_i,a_i;\theta_i\right)\right)^2\right), \qquad (7.1)$$

where

$$y_i=\begin{cases} r_i, & \text{if } s_i' \text{ is terminal condition} \\ r_i+\gamma \max_{a'} Q_i\left(s_i',a_i';\theta_i'\right), & \text{otherwise} \end{cases}.$$

θ_i' represents the parameters of the target network with the same structure as an evaluation network. r_i denotes the one-step reward. The designed training algorithm flowchart is presented in Figure 7.2 to illustrate the planning procedure more clearly. In each network, we adopt a three-layer structure, and the number of neurons in three layers are 5, 100, and 31, respectively. However, there also exist many challenges in training process under our designed networks. For example, due to the parallel network architecture, the update rates of two network parameters in one episode are not synchronized, which greatly increases the convergence difficulty. Besides, since the number, position, and speed of obstacles are unknown, there may be too many situations, which brings many challenges to the design of state space. Consequently, we present two tricks in the training process:

1. In much of the planning literature on learning methods, it is generally designed to stop an episode after reaching the target point due to the static and simple environment. Therefore, considering that the environment we study is more dynamic and changeable, we design the termination condition of an episode as the maximum number of planning steps or reaching the target within the maximum steps. This trick can effectively reduce the useless samples, thereby greatly improving the convergence rate.

2. When we plan the path in a real environment, the obstacle positions need to be detected in real time. However, in order to further improve convergence rate, we suppose that the obstacle positions are available in training process. Besides, this kind of treatment can also improve generalization performance, the main reason for which is more effective samples.

7.3.2 Training Setup

7.3.2.1 State Space

In addition to preceding tricks in the training process, it is crucial to design the suitable state space. First, we define initial position, target position, and current position as $p_o = (x_o, y_o)$, $p_g = (x_g, y_g)$, and $p_c = (x_c, y_c)$, respectively. $p_{obs}^i = (x_{obs}^i, y_{obs}^i)$ denotes the obstacle position, where i represents the obstacle number. According to the goals of path planning, we define five states as follows:

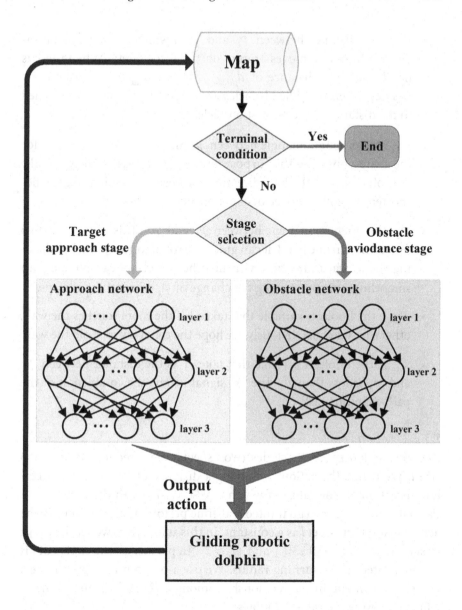

FIGURE 7.2 Framework and flowchart of path planning with HDQN.

- d_{obs}: the distance between p_c and p_{obs}^i, which is used to guarantee that the robot moves around obstacles for safety. When there is no obstacle at a distance of $d_{threshold}$, we set d_{obs} as a constant d_0. Besides, when multiple obstacles are detected, d_{obs} will be assigned to the distance of the nearest obstacle.

- φ_{obs}: the orientation angle between p_c and p_{obs}^i. Similarly, φ_{obs} is for the nearest obstacle. The purpose for designing this feature is to make the robot choose the best direction for obstacles avoidance, further shorten the collision-free distance as much as possible.

- φ_g: the orientation angle between p_c and p_g. This feature has two functions. On one hand, it can guide the robot to goal point by adjusting the forward direction. On the other hand, we can pursue path smoothness via optimizing the change of φ_g in continuous steps.

- B_{wall}: the Boolean variable that takes 1 if the robot touches the wall, otherwise takes 0. Obviously, we hope the robot try to avoid the wall.

- B_{goal}: the Boolean variable that takes 1 as the distance between p_c and p_g is less than δ. The "1" signal means the completion of the path-planning task.

7.3.2.2 Action Space

Actually, action space concludes two aspects: the direction φ and step size l. We define the action as moving a distance of step size l in a certain direction φ, the range of which is $[\varphi_{min}, \varphi_{max}]$ with the interval φ_o. Particularly, in order to guarantee real-time performance, we simplify the action space, that is, set l as a constant. In this subsection, we set the range of action space at $[-30°, 30°]$ and $\varphi_o = 2°$. In particular, the choice of l is closely related to the turning radius. We use a polygon to approximate a circle, and can obtain the relationship among l, φ, and turning radius r using the cosine theorem as follows:

$$l = r\sqrt{2\left(1 - \cos\left(\varphi\right)\right)}. \tag{7.2}$$

Next, we can see that r takes the minimum value when $l = l_{min}$ and $\varphi = \varphi_{max}$. Furthermore, we have analyzed the planar turning maneuverability, which illustrated that the gliding underwater robot could achieve the minimum 0.2-m turning radius. Therefore, by substituting $r_{min} = 0.2$ m and $\varphi_{max} = 30°$, we can calculate $l_{min} = 0.1$ m. In actual training and testing process of HDQN, we set $l = 0.2$ m, which is larger than its minimum value.

7.3.2.3 Reward Function

With full consideration of path smoothness, length, and safety, we formulate the reward function as follows:

$$R = \sum_{i=1}^{6} \omega_i r_i, \tag{7.3}$$

where ω_i represents the positive weight coefficient for the reward items, which is manually set. Regarding r_i, we design them as follows:

- $r_1 = -d_g/d_{max}$, where d_g denotes the distance from the current position to the goal, and d_{max} is set as the diagonal distance of the map for normalization.

- $r_2 = f(d_{obs})$, where $f(d_{obs})$ can be set as inverse proportional, logarithmic, or constant functions. By comparison, we finally select the following function:

$$f(x) = -\frac{1}{2}\left(\frac{1}{x} - \frac{1}{a}\right), \tag{7.4}$$

where a denotes the threshold of the obstacle distance.

- $r_3 = -|\varphi_g|/2\pi$. This item aims to make the robot move toward the target point.

- $r_4 = -g(|e_\varphi|)/\omega_4$, where $e_\varphi = \varphi - \varphi'$ represents the orientation error between the current and previous steps. This item is proposed to smooth the action. In this subsection, we employ the piecewise function as follows:

$$g(x) = \begin{cases} C_0 & x \geq 30° \\ C_1 & 20° \leq x < 30° \\ C_2 & 10° \leq x < 20° \\ C_3 & 5° \leq x < 10° \\ 0 & x < 5°. \end{cases} \tag{7.5}$$

- $r_5 = -C_{wall} / \omega_5$, where C_{wall} can be set as a positive constant if the robot crosses the border of the map, else 0.

- $r_6 = C_{goal} / \omega_6$. When the robot enters the threshold circle of the target point, we assign C_{goal} a large positive reward, else 0.

7.4 PATH-FOLLOWING CONTROL

7.4.1 Problem Formulation and LOS Law

For path following problem of the gliding underwater robot in the horizontal plane, the kinematic equation of the underwater robot can be formulated by

$$\dot{x} = u\cos\psi - v\sin\psi$$
$$\dot{y} = u\sin\psi + v\cos\psi , \qquad (7.6)$$
$$\dot{\psi} = r$$

where (x, y, ψ) represent the real-time position and yaw angle with respect to (w.r.t.) inertia frame, respectively. (u, v, r) denote the line and angular

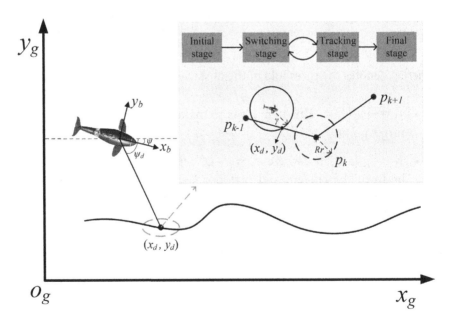

FIGURE 7.3   The illustration of LOS law stages and coordinate systems.

velocities w.r.t. body frame, respectively. Thereafter, the dynamic model can be derived by

$$\dot{u} = \frac{1}{m_{11}}\left(\tau_u + m_{22}vr - d_{11}u\right)$$

$$\dot{v} = \frac{1}{m_{22}}\left(-m_{11}ur - d_{22}v\right) \quad , \qquad (7.7)$$

$$\dot{r} = \frac{1}{m_{33}}\left(\tau_r + (m_{11} - m_{22})uv - d_{33}r\right)$$

where $M = diag(m_{11}, m_{22}, m_{33})$ and $D = diag(d_{11}, d_{22}, d_{33})$ are the mass and damping parameters larger than zero. τ_u and τ_r indicate the thrust force and yaw moment, respectively.

On the basis of earlier planned path, we further apply the LOS guidance law to obtain target yaw angle according to the current position and foresight point. The coordinate framework and LOS system are illustrated in Figure 7.3. $C_g = o_g x_g y_g$ and $C_b = o_b x_b y_b$ represent the inertia and body frames, respectively. Thus, we can define some control variables as follows:

$$\begin{aligned} x_e &= x - x_d \\ y_e &= y - y_d \\ \psi_e &= \psi - \psi_d \\ e_{xy} &= \sqrt{x_e^2 + y_e^2} \, , \end{aligned} \qquad (7.8)$$

where $p_d = (x_d, y_d)$ is desired position of the gliding underwater robot and is obtained by LOS guidance law. ψ_d denotes the desired orientation which can be calculated by

$$\psi_d = \begin{cases} -0.5\pi\,\mathrm{sign}\left(y_e\right), \text{when } x_e = 0 \\ \arctan\left(y_e / x_e\right) + 0.5\pi\,\mathrm{sign}\left(y_e\right)\left[1 + \mathrm{sign}\left(x_e\right)\right], \\ \text{when } x_e \neq 0 \end{cases} \qquad (7.9)$$

where $\mathrm{sign}(\cdot)$ denotes the sign function with $\mathrm{sign}(0)=1$.

With regard to the LOS guidance law, there are four stages including the initial stage S_0, switching stage S_1, tracking stage S_2, and final stage. Furthermore, we design two virtual circles associated with the robot and way point, of which the radii are γ and R_r, respectively. In the initial and final stages, the desired points are selected as the first and last waypoints, respectively. In particular, we improve the calculation in switching stage.

When the robot enters the virtual circle of way point p_k, we choose the desired point as the foresight intersection of the robot virtual circle and line segment $p_k p_{k+1}$ rather than that of the way point virtual circle and $p_k p_{k+1}$. The advantages of this improvement are twofold:

1. The change of the foresight point is gentler so that its derivative $(\dot{x}_d, \dot{y}_d)$ and following error e_{xy} will not change abruptly, which further offers more stable inputs to the controller.

2. The robot can pass the waypoints closer while following the path, which not only ensures path smoothness but also reduces following error to a certain extent.

Since the virtual circle moves in real time as the position of the robot changes, these intersections are also continuously changing. Therefore, through applying the LOS law, we can convert the discrete planning points into continuous desired location points, thus laying the foundation for the subsequent controller design. Regarding the engineering issue of path planning and following for a gliding underwater robot, the general framework is shown in Figure 7.4. For the following control subsystem, we make some basic assumptions as follows:

Assumption 1:

1. The roll and pitch angle is ignored since the path following problem is discussed in the horizontal plane.

2. The reference states $x_d, y_d, \psi_d, \dot{x}_d, \dot{y}_d, \dot{\psi}_d$ are all bounded.

7.4.2 Controller Design

In this subsection, we apply the backstepping methodology to design heading and velocity controllers. The detailed procedures are as follows:

Step 1: The first backstepping state variable ψ_e is chosen as z_1

$$z_1 = \psi_e$$
$$\dot{z}_1 = \dot{\psi} - \dot{\psi}_d = r - \dot{\psi}_d \qquad (7.10)$$

Thus, we define a barrier Lyapunov functions V_1 for z_1 [240][241].

$$V_1 = \frac{1}{2}\log\frac{\psi_L^2}{\psi_L^2 - z_1^2} \qquad (7.11)$$

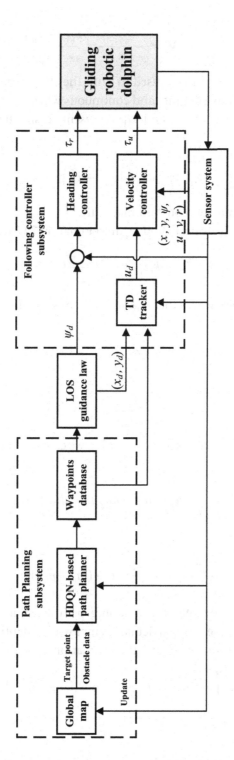

FIGURE 7.4 System framework of path planning and following.

$$\dot{V}_1 = \frac{z_1 \dot{z}_1}{\psi_L^{\,2} - z_1^{\,2}} = \frac{z_1\left(r - \dot{\psi}_d\right)}{\psi_L^{\,2} - z_1^{\,2}}, \tag{7.12}$$

where ψ_L is a positive constant used to limit the state variable. It can be shown that V_1 is positive definite and continuous when $|z_1| < \psi_L$.

Then, we define $r_e = r - \alpha_1$, and α_1 represents a stabilizing function designed as follows:

$$\alpha_1 = \dot{\psi}_d - k_1 z_1 \left(\psi_L^{\,2} - z_1^{\,2}\right), \tag{7.13}$$

where k_1 is set as a positive constant manually.

Therefore, by substituting r_e and α_1 into Equation 7.12 we can yield

$$\dot{V}_1 = -k_1 z_1^{\,2} + \frac{z_1 r_e}{\psi_L^{\,2} - z_1^{\,2}}. \tag{7.14}$$

Step 2: In this step, we further stabilize r_e; thus, we define the second Lyapunov functions

$$V_2 = V_1 + \frac{1}{2} r_e^{\,2} \tag{7.15}$$

$$\dot{V}_2 = -k_1 z_1^{\,2} + \frac{z_1 r_e}{\psi_L^{\,2} - z_1^{\,2}} + r_e(\dot{r} - \dot{\alpha}_1). \tag{7.16}$$

According to the dynamics of yaw angular, we can formalize the derivative of r_e as follows:

$$\dot{r}_e = \dot{r} - \dot{\alpha}_1 = \frac{1}{m_{33}}\left(\tau_r + (m_{11} - m_{22})uv - d_{33}r\right) - \dot{\alpha}_1. \tag{7.17}$$

Thus, we can obtain

$$\dot{V}_2 = -k_1 z_1^{\,2} - r_e \dot{\alpha}_1 + \frac{z_1 r_e}{\psi_L^{\,2} - z_1^{\,2}} + r_e\left(\frac{1}{m_{33}}\left(\tau_r + (m_{11} - m_{22})uv - d_{33}r\right)\right). \tag{7.18}$$

Step 3: Since d_{33} denotes the damping parameter, it is difficult to acquire its precise value. We particularly employ the adaptive methodology to estimate it. Thus, we define the estimation of d_{33} as $\hat{d}_{33}$, and its error variable $\tilde{d}_{33} = d_{33} - \hat{d}_{33}$.

Assumption 2: The parameter d_{33} will not change in a short time; that is, $\dot{d}_{33} = 0$.

Next, we define the third Lyapunov function:

$$V_3 = V_2 + \frac{1}{2}\tilde{d}_{33}^{\ 2}$$ (7.19)

$$\dot{V}_3 = -k_1 z_1^{\ 2} - \tilde{d}_{33}\dot{\hat{d}}_{33} + \frac{z_1 r_e}{\psi_L^{\ 2} - z_1^{\ 2}} + r_e$$ (7.20)

$$\left(\frac{1}{m_{33}}(\tau_r + (m_{11} - m_{22})uv - d_{33}r) - \dot{\alpha}_1 \right).$$

Then, we design the control law τ_r as follows:

$$\tau_r = (m_{22} - m_{11})uv + \hat{d}_{33}r + m_{33}\left(\dot{\alpha}_1 - k_2 r_e - \frac{z_1}{\psi_L^{\ 2} - z_1^{\ 2}} \right),$$ (7.21)

where k_2 is set as a positive constant manually.

By substituting the designed control law, we can obtain

$$\dot{V}_3 = -k_1 z_1^{\ 2} - k_2 r_e^{\ 2} - \tilde{d}_{33}\left(\frac{r_e r}{m_{33}} + \dot{\hat{d}}_{33} \right).$$ (7.22)

Therefore, we regard the $\dot{\hat{d}}_{33} = -\frac{r_e r}{m_{33}}$, i.e., $\hat{d}_{33} = -\frac{1}{m_{33}}\int_0^t r_e r dt$, and yield

$$\dot{V}_3 = -k_1 z_1^{\ 2} - k_2 r_e^{\ 2} \le -\lambda V_2 \le -\lambda V_3,$$ (7.23)

where $\lambda = \min\{k_1, k_2\}$.

Thus, we rearrange the control law as follows:

$$\tau_r = (m_{22} - m_{11})uv - \frac{1}{m_{33}}\int_0^t r_e r dt + m_{33}\left(\dot{\alpha}_1 - k_2 r_e - \frac{z_1}{\psi_L^{\ 2} - z_1^{\ 2}} \right).$$ (7.24)

Lemma 1:

1. $V > 0$;

2. $\dot{V} \le -g(t)$, where $g(t) \ge 0$;

3. if $\dot{g}(t)$ is bounded, and $g(t)$ is uniformly continuous, then $\lim_{t \to \infty} g(t) = 0$.

Remark 1: Apparently, $V_3 > 0$, and we regard $g(t) = -k_1 z_1^2 - k_2 r_e^2$. Obviously, $g(t)$ satisfies Lemma 1. Hence, we can conclude that $\lim_{t \to \infty} \left(k_1 z_1^2 + k_2 r_e^2\right) = 0$, thereby proving the orientation error under our control law will converge to zero.

Remark 2: According to lemma in [240], we can know $|z_1| = |\psi_e| < k_{b1} = \psi_L, \forall t \in [0, \infty)$, while providing that the initial conditions $z_1(0) < k_{b1}$.

Step 4: According to the LOS guidance law, we can see that $e_{xy} \le \sigma$, and σ is a small positive constant. Thus, the purpose of path following is to let the tracking error $(e_{xy} - \sigma) \to 0$, and we can define the Lyapunov functions

$$V_4 = V_3 + \frac{1}{2}\left(e_{xy} - \sigma\right)^2. \qquad (7.25)$$

On the basis of kinematics, we can derive that

$$\dot{e}_{xy} = -u\cos\psi_e + v\sin\psi_e - \left(\dot{x}_d \cos\psi_d + \dot{y}_d \sin\psi_d\right). \qquad (7.26)$$

Then, the derivative of V_3 can be formalized as follows:

$$\dot{V}_4 = \dot{V}_3 + \left(e_{xy} - \sigma\right)\dot{e}_{xy}. \qquad (7.27)$$

Next, we define the velocity error as $u_e = u - \alpha_2$, and design the virtual input as follows:

$$\begin{aligned}\alpha_2 &= \cos^{-1}\left(\psi_e\right)\left(k_3\left(e_{xy} - \sigma\right) + v\sin\left(\psi_e\right)\right), \\ &\quad - \cos^{-1}\left(\psi_e\right)\left(\dot{x}_d \cos\psi_d + \dot{y}_d \sin\psi_d\right)\end{aligned} \qquad (7.28)$$

where k_3 is as a positive constant.

Thus, we can obtain

$$\begin{aligned}\dot{V}_4 &= \dot{V}_3 + \left(e_{xy} - \sigma\right)\left(-u_e\cos\left(\psi_e\right) - k_3\left(e_{xy} - \sigma\right)\right) \\ &= -k_1 z_1^2 - k_2 r_e^2 - k_3\left(e_{xy} - \sigma\right)^2 - u_e\cos\left(\psi_e\right)\left(e_{xy} - \sigma\right).\end{aligned} \qquad (7.29)$$

Remark 3: With the aid of the designed barrier Lyapunov functions and Lemma 2, we can guarantee the error of yaw angle $|\psi_e| < \pi/2$, via setting $\psi_L = \pi/2$ and $\psi_e(0) < \pi/2$. Therefore, the singularity of the $\cos^{-1}(\psi_e)$ can be avoided. In particular, this kind of treatment is consistent with our path-planning idea. If $\psi_e > \pi/2$, we close the controller, and adjust the attitude using the flippers in an open-loop way until $\psi_e < \pi/2$. Since the turning radius of the gliding underwater robot is constrained, this improvement can significantly reduce the length of following path while sacrificing some smoothness.

Step 5: Afterward, we stabilize the dynamics of u_e by designing the Lyapunov function

$$V_5 = V_4 + \frac{1}{2}u_e^2. \tag{7.30}$$

According to the dynamics of the forward velocity, we can formalize the derivative of V_5 as follows:

$$\dot{V}_5 = \dot{V}_4 + u_e \dot{u}_e = \dot{V}_4 + u_e \left(\dot{u} - \dot{\alpha}_2 \right)$$

$$= \dot{V}_4 + \frac{u_e}{m_{11}} \left(\tau_u + m_{22} vr - d_{11} u \right) - \dot{\alpha}_2 u_e. \tag{7.31}$$

In the same way, we also apply the adaptive method to estimate the damping parameter d_{11} by defining $\tilde{d}_{11} = d_{11} - \hat{d}_{11}$, and Assumption 1 is also applicable to d_{11}. Thus, we design the control law as follows:

$$\tau_u = -m_{22} vr + \hat{d}_{11} u + m_{11} \left(\dot{\alpha}_2 - k_4 u_e + \cos(\psi_e)(e_{xy} - \sigma) \right), \tag{7.32}$$

where

$$\hat{d}_{11} = -\frac{1}{m_{11}} \int_0^t u_e u \, dt$$

k_4 is a positive constant.

Therefore, the derivative of v_5 can be derived by

$$\dot{V}_5 = -k_1 z_1^2 - k_2 r_e^2 - k_3 \left(e_{xy} - \sigma \right)^2 - k_4 u_e^2. \tag{7.33}$$

Similarly, according to Lemma 1, we can conclude that

$$\lim_{t \to \infty} \left(k_1 z_1^2 + k_2 r_e^2 + k_3 (e_{xy} - \sigma)^2 + k_4 u_e^2 \right) = 0,$$

which indicates that the following error under our control law will converge to zero. With regard to the bound of the following error, we further make some analysis. First, in velocity controller design, we define the Lyapunov function as follows:

$$V = \frac{1}{2}e^2 + \frac{1}{2}u_e^2, \tag{7.34}$$

where $e = e_{xy} - \sigma$. Next, we can also calculate the dot of e and obtain the relationship between e and u as follows:

$$u = \frac{-\dot{e} + v\sin\psi_e - \left(\dot{x}_d\cos\psi_d + \dot{y}_d\sin\psi_d\right)}{\cos(\psi_e)}. \tag{7.35}$$

Furthermore, we can obtain $u_e = \cos^{-1}(\psi_e)(-\dot{e} - k_3 e)$. Next, by substituting our control law, we can obtain the $\dot{V} = -k_3 e^2 - k_4 u_e^2$. Thus, we can calculate the final form of $\dot{V}$ as follows:

$$\dot{V} = \begin{pmatrix} e & \dot{e} \end{pmatrix} M \begin{pmatrix} e \\ \dot{e} \end{pmatrix} \tag{7.36}$$

$$M = -\frac{1}{\cos^2(\psi_e)} \begin{pmatrix} k_3\cos^2(\psi_e) + k_3^2 k_4 & k_3 k_4 \\ k_3 k_4 & k_4 \end{pmatrix}. \tag{7.37}$$

Due to $\dot{V} < 0$, and $V(e,\dot{e}) < V(0)$, we can identify that the vector $z = (e,\dot{e})^T$ is bounded according to Lyapunov's Theorem [242], and $\|z\|^2 \leq \beta\|z(0)\|^2$, where β is a positive constant.

Furthermore, since the desired velocity α_2 and the derivative of the desired position $(\dot{x}_d, \dot{y}_d)$ are susceptible to environmental disturbances, we employ the tracking differentiator for tracking and filtering them for engineering treatment.

7.5 SIMULATIONS AND EXPERIMENTS

In this section, extensive typical simulations and aquatic experiments are conducted to validate the performance of the proposed path planner as well as the effectiveness of the nonlinear adaptive backstepping controller for the path following.

7.5.1 Results of Path Planning

In path-planning simulations, we first built up the environment. This environment includes two kinds of information: map and obstacle information. Referring to the size of the real testing pool, the virtual pool was designed with the dimension of 4 m long and 4 m wide. Second, the obstacles can be randomly arranged in certain positions, but we set the maximum number of obstacles to six in view of the size of the pool. Besides, we set the radius of obstacles to be fixed at 0.3 m.

Figure 7.5 indicates the training results under the net of natural DQN, double DQN (DDQN), prioritized replay DQN (DPQN), deep recurrent Q-network (DRQN), and proposed HDQN. Note that we adopted the same training conditions in a comparative training process including some key parameters, such as learning rate, discount factor, and total episodes. These parameters of HDQN are tabulated in Table 7.1. From Figure 7.5, we found that the rewards of DQN, DPQN, and HDQN converged to zero, while DDQN and DRQN showed poor performances. In particular, since the net structure of DRQN and HDQN were different from other networks, resulting in different calculation criteria of the reward value, their initial training rewards were lower. However, as the training progresses, the reward curve of HDQN gradually caught up with DQN and DPQN, which meant the reward promotion of HDQN was better than the other techniques. More important, we also analyzed the training times, which were 2.17 h, 2.17 h, 7.92 h, 2.25 h, and 1.3 h for DQN, DDQN, DPQN, DRQN, and HDQN, respectively. Regarding HDQN with the shortest training time,

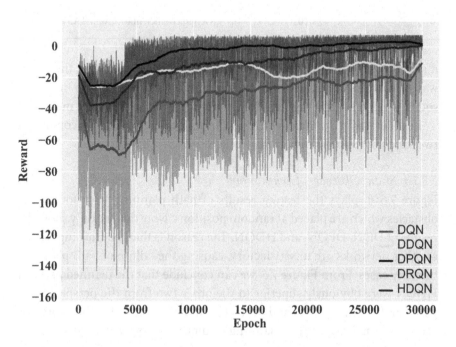

FIGURE 7.5 The training results under different networks.

TABLE 7.1 Parameters of HDQN

Parameters	Value	Meaning
lr	0.00001	Learning rate
γ	0.9	Discount factor
N_{update}	100	Update frequency of the target network
N_{replay}	200000	Capacity of replay memory
N_{batch}	256	Number of sampled transition
N_{epoch}	30000	Number of explore episodes
$w_{1-app} \mid w_{1-obs}$	0.1|0.01	Weight coefficient of r_1 w.r.t TAS and OAS
$w_{2-app} \mid w_{2-obs}$	0.01|1	Weight coefficient of r_2 w.r.t TAS and OAS
$w_{3-app} \mid w_{3-obs}$	1.5|0.1	Weight coefficient of r_3 w.r.t TAS and OAS
$C_{0|1|2|3-app}$	0.01|0.005|0.0025|0	Weight coefficient of r_4 w.r.t OAS
$C_{wall-app} \mid C_{wall-obs}$	0.5|0.5	Weight coefficient of r_5 w.r.t TAS and OAS
$C_{goal-app} \mid C_{goal-obs}$	10|10	Weight coefficient of r_6 w.r.t TAS and OAS

it meant that in the middle and later stages of training process, the target was successfully reached within the maximum training steps in most episodes, which also demonstrated the superiority of the proposed method.

In order to verify the effectiveness of trained networks, we conducted two kinds of testing simulations.

7.5.1.1 Static Obstacles Environment

Figure 7.6 displays the testing results of path planning with four static obstacles which are placed at random positions. Note that we only show the results of DQN, DPQN, and HDQN. The reason is that the training results of other networks are unsatisfactory, causing the failure of path planning in many cases. From Figure 7.6, we can conclude that the planned paths of HDQN were obviously superior to the other two from the perspective of path length, smoothness, and obstacle avoidance safety. Furthermore, for the sake of quantifying the planning results, we developed a test standard

$$Penalty = p_{length} + p_{smooth} + p_{obstacle} + p_{wall}\cdot, \tag{7.38}$$

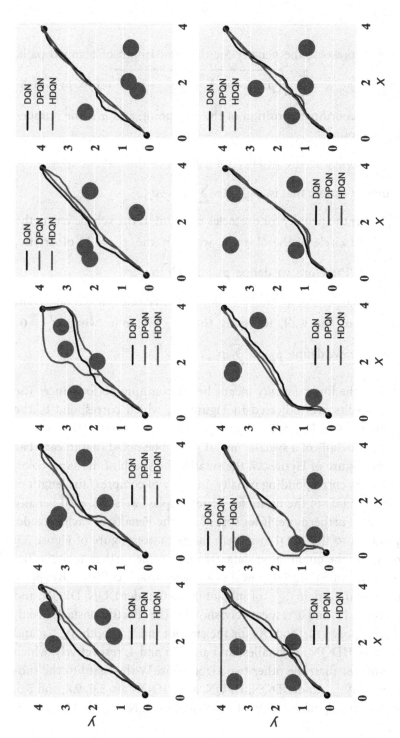

FIGURE 7.6 The path illustration of testing results compared with natural DQN and DPQN.

where

- P_{length} represents the penalty for the total length of planned path, that is, $P_{length} = \sum_{i=1}^{n-1} \sqrt{\left(px_{i+1} - px_i\right)^2 + \left(py_{i+1} - py_i\right)^2}$. (px, py) presents the coordinate position of the way point, and n is the number of path points.

- P_{smooth} indicates the penalty for path smoothness and adopts the change of action, that is, $P_{smooth} = \sum_{i=1}^{n-1} |a_{i+1} - a_i|$.

- $P_{obstacle}$ is the penalty for obstacles collision. If the vehicle enters the threshold circle of the obstacle, we apply the $p_{obs}^i = 4$; otherwise, $p_{obs}^i = 0$. Therefore, we define $P_{obstacle} = \sum_{i=1}^{n} p_{obs}^i$.

- P_{wall} is the penalty for wall collision. If the vehicle is less than 0.04 m away from the wall, we apply the $p_{wa}^i = 4$; otherwise, $p_{wa}^i = 0$. Therefore, we define $P_{wall} = \sum_{i=1}^{n} p_{wa}^i$.

Therefore, the lower *Penalty* means better planning performance. The quantized results are indicated in Figure 7.7, which correspond to the testing situations in Figure 7.6. Figure 7.7 shows the penalty scores of 10 cases and the details of a specific one. It should be noted that in each bar of the upper figure of Figure 7.7, the smaller the width of the same color, the smaller the corresponding penalty. In order to compress the length of the horizontal axis of the figure for better display, we stack each penalty item to show. Furthermore, if we spread out the Penalty of each episode, it is equivalent to the "total loss" item in the bottom figure of Figure 7.7. With regard to the quantitative data, we make some further analysis. The indexes 0 and 2 of Figure 7.6 show the obvious obstacle collision. By applying our test standard, $P_{obstacle}$ of subplot index 0 under DQN, DPQN, and HDQN are 12, 12, and 0, respectively, showing the effective obstacle avoidance of a proposed method. As for the subplot index 2, both P_{smooth} and $P_{obstacle}$ under HDQN are smallest and are 5.75 and 4, respectively, which is much smaller than the other two algorithms. With regard to the subplot index 5, P_{length} under DQN, DPQN, and HDQN are 5.6, 9.8, and 5.6, respectively. P_{smooth} under DQN, DPQN, and HDQN are 9.2, 14.3, and 4,

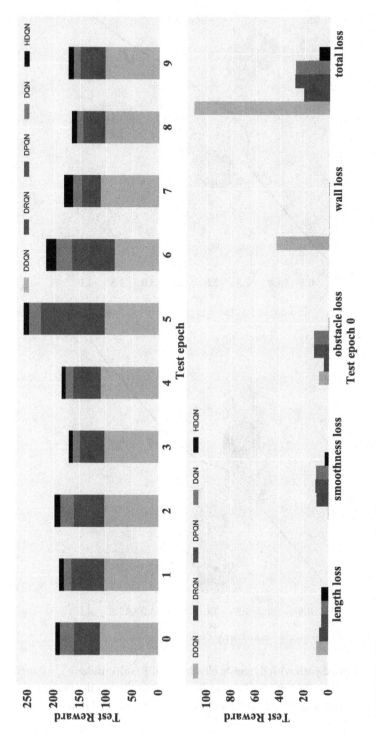

FIGURE 7.7 The comparative penalty illustration of testing results.

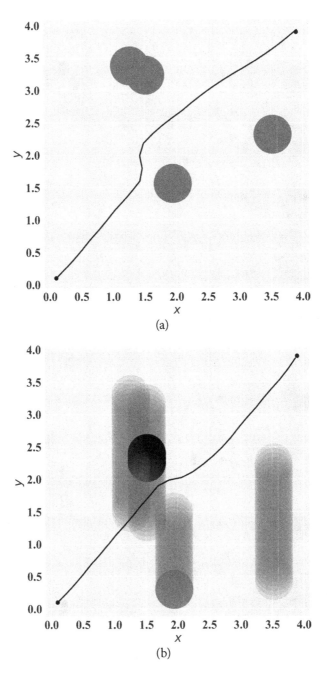

FIGURE 7.8   Planned path with dynamic obstacles. (a) Results under $v_{obs} = 0$ m/s. (b) Results under $v_{obs} = 0.07$ m/s and down direction flow. (c) Results under $v_{obs} = 0.1$ m/s and left direction flow. (d) Results under $v_{obs} = 0.13$ m/s and random direction flow.

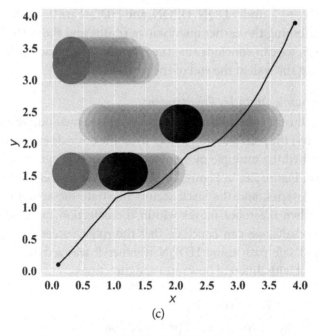

(c)

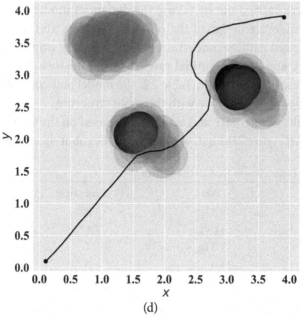

(d)

FIGURE 7.8 *(Continued)*

respectively. p_{wall} under DQN, DPQN, and HDQN are 0, 17, and 0, respectively. Combining the earlier quantitative results and the path in the subplot index 5, we can conclude that the robot does not reach the target and moves along the wall at the end of the path.

7.5.1.2 Dynamic Obstacles Environment

To further illustrate the advantages of the proposed method, we also applied HDQN to test the path planning under dynamic obstacle environments and full researched the multiple movement directions and speeds of obstacles, as shown in Figure 7.8. In Figure 7.8, the red sectors represent the dynamic range of obstacles, and the black sectors illustrate the actual positions of obstacles when the robot moves within the detection range of obstacles. From the results, we can conclude that the robot successfully planned a smooth and safe path using HDQN whether it was a dynamic obstacles environment with different flow rates and directions.

7.5.2 Results of Path Following

To illustrate the performance of proposed control method, extensive simulations were carried out by taking the dynamics model in Equation 7.7. First, we randomly designed the obstacle locations and used HDQN to plan the desired path. Second, the robot started at the origin of the coordinate system, and both the initial yaw angle and the forward velocity were zero. Besides, as tabulated in Table 7.2, the model and control parameters are quantified in accordance with the robot structure.

The results of reference and actual paths based on the initial conditions are plotted in Figure 7.9. First, we found that the robot successfully followed

TABLE 7.2 Parameters of the Control System

Parameters	Value	Parameters	Value
m_{11}	9.9 kg	σ	0.02 m
m_{22}	14.5 kg	R_r	0.2 m
m_{33}	1.3 kg	γ	0.2 m
k_1	1.0	δ_{0-x_d}	30
k_2	2.0	δ_{0-y_d}	30
k_3	1.0	δ_{0-u_d}	12
k_4	5.0	h_{0-x_d}	0.3
h_{0-yd}	0.3	h_{0-u_d}	0.9

the planned path even for curves with irregular curvature changes. Second, compared with PID and SMC, the proposed method showed better performance. PID's results overshoot while tracking corners, and SMC's showed insufficient tracking ability. In some engineering applications, overshoot characteristic would be considered adverse performance since it can often cause damage to the actuators. In contrast, the robot using the proposed method not only were closer to the target path when tracking the straight line but also performed smaller following error in a smooth manner while tracking a curved segment. Third, we also made some quantitative investigations of the following error that represented the distance between the current point and the tracking line segment. The maximum errors of PID, SMC, and the proposed method were 9.7 cm, 11.96 cm, and 9.7 cm, respectively. Meanwhile, the mean absolute error (MAE) of PID, SMC, and the proposed method were 2.15 cm, 2.71 cm, and 1.99 cm, respectively. Moreover, the yaw angle and velocity information are presented in Figure 7.10. It was indicated that yaw angle and forward velocity tracked the target value successfully, and the sway velocity converged to zero.

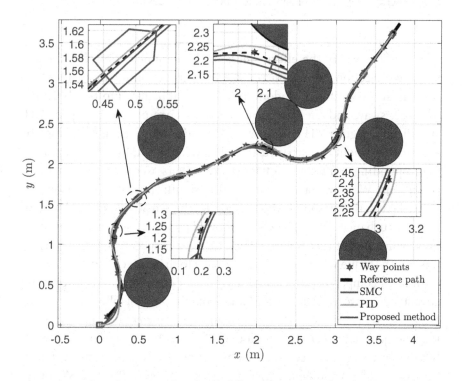

FIGURE 7.9 Simulation results of path following compared with PID and SMC.

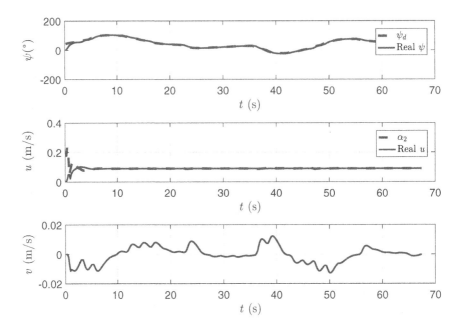

FIGURE 7.10 The results of yaw angle and velocity tracking.

To further verify the effectiveness of the designed controller for disturbance rejection, we conducted the interference simulations, in which the zero mean uniform random noises were affiliated into the dynamics of surge, sway, and the yaw axis. The results are plotted in Figure 7.11. Specifically, Figure 7.11a illustrates the path comparison including the ABP with/without disturbance and BP, SMC, and PID with disturbance. Corresponding parameters of these controllers were same as previous conditions. We saw that the path of PID significantly deviated from the desired path, the reason of which might be the sensitivity and vulnerability to disturbance. The path of SMC showed poor performance in smoothness, which meant that the calculated torque changed frequently, whereas unsmooth torque will lead to great damage to drive mechanism.

In brief, compared with existing techniques, our proposed controller has advantages in two aspects. On one hand, from the point of following error, the proposed controller has better performance. In static environment, it seems that the quantitative error indicators of three methods are relatively close, but the differences between their paths can be clearly seen from the figure. In both Figure 7.9 and Figure 7.11a, the performances of PID and SMC have overshoot characteristics and insufficient following ability in the turning

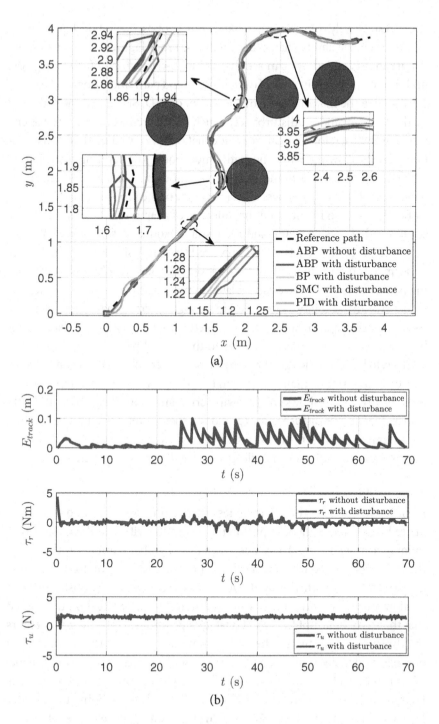

FIGURE 7.11 Simulation results of path following with external disturbance compared with other control methods. (a) Results of the desired path and the actual path. (b) Results of calculated propulsive force and moment.

stage, respectively. On the other hand, our proposed method has stronger anti-interference ability than other methods. From Figure 7.11, once random disturbance is added, we can see that the following effects of PID and SMC decline more severely in view of overshooting and path smoothness. Moreover, we also give the maximum following errors of PID, SMC, and the proposed method under random disturbance, which are 12.4 cm, 14.1 cm, and 9.8 cm, respectively. Therefore, we can conclude that our proposed method still maintains small following error and path smoothness after adding disturbance, while the performances of the other two methods have declined significantly, which reveals the effectiveness and superiority of the proposed methods.

Besides, we also investigated the influence of adaptive parameters. Thus, we set d_{11} and d_{33} as constant, and the result is plotted in the legend "BP with disturbance" of Figure 7.11a. It was revealed that an ABP controller had better performance than a single BP. Since we adopt the adaptive technique for damping parameters, which are closely related to mass parameters, it is beneficial to reduce system instability factors. If there are slightly wrong mass parameters, the following performance of the proposed method will not cause very bad results, because it is not particularly sensitive to the control parameters. Therefore, it is suggested that the designed controller has better robustness against disturbances, and it may be mainly owing to the adaptively tuning damping parameters. Figure 7.11(b) shows the following error and control signals of the ABP results. Due to the disturbance, τ_r and τ_u showed intensive undulation in a certain range. Besides, the following error was basically equivalent to that without disturbance, indicating the effectiveness of the proposed methods.

7.5.3 Experimental Results and Analysis

In order to validate the effectiveness of the proposed methods, we conducted extensive aquatic experiments in a pool with the dimension of 5 m long, 4 m wide, and 1.1 m deep. Besides, a global camera on the top of the pool was applied to offer the real-time map information for the gliding underwater robot. In the pool, we set four 0.15-m-radius obstacles with random positions and set the target position at (3.8, 3.8). It should be noted that the robot does not know the obstacles positions. Until the robot enters the threshold of obstacles, the global camera will notify the robot that an obstacle is detected, and then the robot will plan the path in real time. With full consideration of real time for the path planner, we achieved the HDQN in C++ platform with libtorch and tested the running time under Intel Core i5-7200 CPU @2.5GHz. Through running the codes 100 times and averaging, the test running time was 40 ms, which fully met the real-time requirement in path planning.

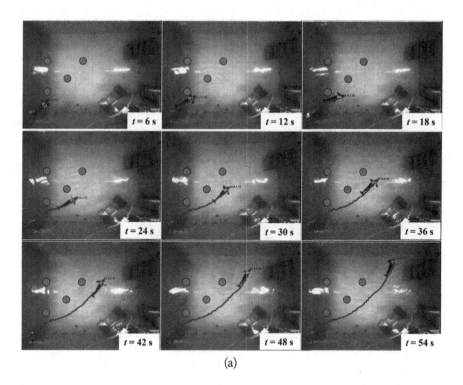

(a)

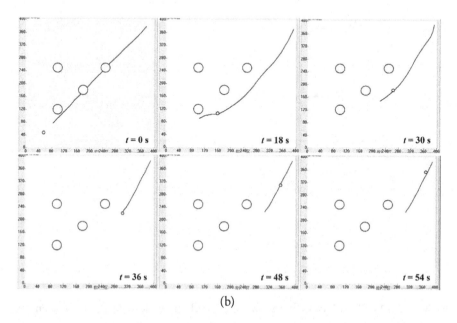

(b)

FIGURE 7.12 The snapshot sequence of the planning and following task. (a) Global camera perspective. (b) Planning and following results.

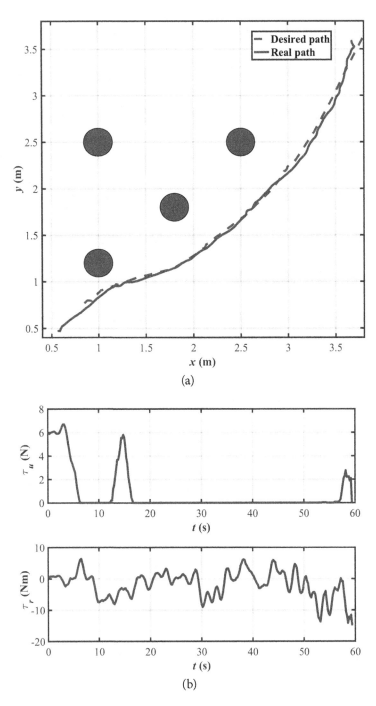

FIGURE 7.13 Results of the following control. (a) Following path. (b) Control force and torque.

Figure 7.12 presents the snapshot sequence of the planning and following task. From Figure 7.12(b), since the obstacles' positions were unknown, we can see that the initial path planning only considered the path length and smoothness. Once the obstacles were detected, the HDQN-based planner worked and planned a satisfactory path with full consideration of path length, smoothness, and safety. It should be noted that in order to show the effect of the following controller, the real-time planner works only when there are obstacles around the robot. Besides, the results of following control are plotted in Figure 7.13a, which indicates that the gliding underwater robot can successfully follow the path. Moreover, since we set the robot virtual circle at 0.2-m radius in LOS law, e_{xy} keeps 0.2 m in the following process, which signifies that the virtual circle of the robot always intersects with the line segment, and further indicating successful following. From Figure 7.13b, we can see that the forward force keeps to zero most of time, the reason for which is that the additional forward velocity generated by turning motion has exceeded the expected speed. From these results, we can conclude that the robot can successfully accomplish the planning and following task in different environments, which indicates that our proposed methods are effective.

7.5.4 Discussion

To efficiently accomplish many ocean tasks for practical searching and cursing applications, we present a hierarchical framework for the gliding underwater robot to solve the issues of poor environmental adaptability and unsatisfactory following control effect. By combining with the designed controller, we successfully accomplish a precise and robust path following task. Compared with traditional path planning, such as methods used by Pêtrès et al. [227], Aghababa [228], Ataei and Yousefi-Koma [229], and Ma et al. [230], the proposed planner shows more superiority in adapting to unknown environment and real-time dynamic planning capabilities. Besides, more complex environments with multiple obstacles are particularly investigated in this study than previous learning-based planning methods. In contrast with path following in Sun et al. [237], Liu [238], and Jia et al. [239], we do not neglect the derivative term of LOS point and avoid the yaw singularities in velocity law, which means that the proposed controller will be more practical in engineering applications.

Since we set the space environment to be continuous, it can be considered that there exist infinite states. Compared with the single network, the convergence rate of HDQN needs to be improved. To address this problem, we can divide the environment into multiple areas, and place the obstacles randomly and sparsely. With regard to the additional speed, it is uncontrollable due to the coupling effect generated by multiple motion modes, which is a disadvantage of the gliding underwater robot. To solve this issue, we intend to improve the mechanical design and add the deceleration mechanism.

7.6 CONCLUSION

In this chapter, a hierarchical engineering framework including the HDQN-based planner and ABP-based controller has been developed to address the issues of path planning and following for the gliding underwater robot. First, the novel planner with HDQN can generate the real-time, smooth, and safe path in both static and dynamic environments, thereby strengthening the environmental adaptability. Second, we improved the LOS law and derived the control law using a backstepping methodology, in which the BLF and damping parameters adaptation are particularly employed to solve the yaw singularities problem and improve robustness with disturbance, respectively. Finally, extensive simulation and experiments not only verified the effectiveness of the proposed planner but also indicated the outperformance of the designed controller. Remarkably, the obtained results offer valuable insight into performing complex ocean exploration using gliding underwater robots.

REFERENCES

[227] C. Pêtrès, Y. Pailhas, P. Patrón, et al., "Path planning for autonomous underwater vehicles," *IEEE Trans. Robot.*, vol. 23, no. 2, pp, 331–341, 2007.

[228] M. P. Aghababa, "3D path planning for underwater vehicles using five evolutionary optimization algorithms avoiding static and energetic obstacles," *Appl. Ocean Res.*, vol. 38, pp. 48–62, 2012.

[229] M. Ataei and A. Yousefi-Koma, "Three-dimensional optimal path planning for waypoint guidance of an autonomous underwater vehicle," *Robot. Auton. Syst.*, vol. 67, pp. 23–32, 2015.

[230] Ma, Y., Y. Gong, C. Xiao, et al., "Path planning for autonomous underwater vehicles: An ant colony algorithm incorporating alarm pheromone," *IEEE Trans. Veh. Technol.*, vol. 68, no. 1, pp. 141–154, 2019.

[231] G. Han, Z. Zhou, T. Zhang, et al., "Ant-colony-based complete-coverage path-planning algorithm for underwater gliders in ocean areas with thermoclines," *IEEE Trans. Veh. Technol.*, vol. 69, no. 8, pp. 8959–8971, 2020.

[232] A. B. Martínez, R. F. Rosa, and F. L. Martínez, "Hierarchical reinforcement learning approach for motion planning in mobile robotics," *IEEE Lat. Am. Rob. Symp.*, 2013, pp. 83–88.

[233] Y. Hu, D. Li, Y. He, et al., "Incremental learning framework for autonomous robots based on Q-learning and the adaptive kernel linear model," *IEEE Trans. Cogn. Develop. Syst.*, vol. 14, no. 1, pp. 64–74, 2019.

[234] E. Fredriksen and K. Y. Pettersen, "Global κ –exponential waypoint maneuvering of ships: Theory and experiments," *Automatica*, vol. 42, no. 4, pp. 677–687, 2006.

[235] B. Li, L. Qiao, and W. Zhang, "Two-time scale path following of underactuated marine surface vessels: Design and stability analysis using singular perturbation methods," *Ocean Eng.*, vol. 124, pp. 287–297, 2016.

[236] X. Liu, M. Zhang, and E. Rogers, "Trajectory tracking control for autonomous underwater vehicles based on fuzzy re-planning of a local desired trajectory," *IEEE Trans. Veh. Technol.*, vol. 68, no. 12, pp. 11657–11667, 2019.

[237] Z. Sun, G. Zhang, B. Yi, et al., "Practical proportional integral sliding mode control for underactuated surface ships in the fields of marine practice," *Ocean Eng.*, vol. 142, pp. 217–223, 2017.

[238] Z. Liu, "Practical backstepping control for underactuated ship path following associated with disturbances," *IET Intell. Transp. Syst.*, vol. 13, no. 5, pp. 834–840, 2019.

[239] Z. Jia, Z. Hu, and W. Zhang, "Adaptive output-feedback control with prescribed performance for trajectory tracking of underactuated surface vessels," *ISA Trans.*, vol. 95, pp. 18–26, 2019.

[240] K. P. Tee, S. S. Ge, and E. H. Tay, "Barrier Lyapunov functions for the control of output-constrained nonlinear systems," *Automatica*, vol. 45, no. 4, pp. 918–927, 2009.

[241] W. He, Y. Chen, and Z. Yin, "Adaptive neural network control of an uncertain robot with full-state constraints," *IEEE Trans. Cybern.*, vol. 46, no. 3, pp. 620–629, 2016.

[242] W. M. Haddad and V. S. Chellaboina, "Nonlinear dynamical systems and control: A Lyapunov-based approach," Princeton University, 2011.

CHAPTER 8

3-D Maneuverability Analysis and Path Planning for Gliding Underwater Robots

8.1 INTRODUCTION

The ocean, which occupies about 71% of the Earth's surface, contains abundant resources. In recent years, ocean development represented by seabed survey, sampling, and seabed operations has attracted the attention of many scientific researchers. As an important part of the preceding tasks, three-dimensional (3-D) path planning in the ocean environment is an essential key technology. However, humans have just explored only about 5% of whole ocean, since the complex underwater environment has brought great challenges to ocean research. The gliding underwater robot has the characteristics of strong endurance and high maneuverability, showing great potential in the field of ocean exploration, but its multi-modal characteristics also put forward higher requirements for 3-D path planning research.

Considering the complex ocean environments, the planned path should be concerned with energy consumption, obstacle avoidance, and path smoothness. In general, many path-planning and -following methods are designed to accomplish underwater tasks for autonomous underwater

DOI: 10.1201/9781003347439-8

vehicles (AUVs) [243]–[244]. Hernandez *et al.* presented an online method for planning collision-free paths for AUVs and used the transition-based rapidly exploring random tree (RRT) algorithm to obtain the path [245]. Yang *et al.* presented a 3-D path-planning approach based on the RRTs and cubic Bezier spiral curves [246]. Wu *et al.* presented path planning for underwater gliders with the artificial potential field approach and added motion constraints into path generation [247], but the generated path was not smooth and largely deviated from the previous track. With regard to path planning for the gliding underwater robot, it can perform more complex tasks due to the integration of gliding and dolphin-like motions. By switching the two motions, the gliding underwater robot can achieve goals, including both low energy consumption and high maneuverability. In addition, to meet the kinematic requirements of the robot, it is particularly important to analyze the motion characteristics and then obtain the motion constraints mainly including the turning radius and attitude Euler angles.

The research on the horizontal path planning of the gliding underwater robot has been introduced in Chapter 7, which is mainly oriented to the relatively narrow underwater environment and completes the task by giving full play to the motion characteristics of the dolphin-like motion. However, for the 3-D path-planning problem in the ocean environment, the spatial scale is larger, and both high maneuverability obstacle avoidance and energy consumption should be considered [248]. Therefore, how to use the strong endurance and high maneuverability of the gliding underwater robot to design a 3-D path planning framework suitable for the marine environment is a question worthy of discussion. In addition, during the previous horizontal path-tracking experiments, it was found that the yaw maneuvering performance of the gliding underwater robot was still insufficient. Although the steering radius is small, the steering angular velocity is limited. Therefore, how to improve the yaw maneuverability and redesign the mode is an urgent problem to be solved.

Further progress is made in this chapter, whose objective is to design a 3-D path-planning framework suitable for a gliding underwater robot. The contributions made in this chapter are summarized as follows:

- An improved yaw mechanism is designed with a rudder so that the yaw maneuverability is greatly strengthened. The yaw and pitch maneuver strategies are proposed, and a 3-D maneuverability analysis is performed. The obtained results can provide kinematic constraints for 3-D path planning.

- A multimodal 3-D path planning framework for the gliding underwater robot in the marine environment is proposed; that is, the gliding motion is used to save energy when cruising while the dolphin-like motion is used to obtain high maneuverability during obstacle avoidance. In detail, the main contents include glide path planning based on geometric constraints, obstacle-avoidance path planning based on dolphin-like motion, and a path-smoothing algorithm.

- The three-dimensional maneuvering characteristics of the developed platform are analyzed through underwater experiments, and the effectiveness of the proposed planning method is also verified in the simulation results.

The remainder of this chapter is organized as follows: Section 8.2 provides the 3-D maneuverability analysis followed by the experimental results. Section 8.3 presents a 3-D path-planning framework. Section 8.4 shows the simulation results followed by a discussion. Section 8.5 concludes this chapter with an outline of future work.

8.2 3-D MANEUVERABILITY ANALYSIS

8.2.1 Horizontal Motion

In this section, the belly rudder is added to the developed platform to explore the degree of improvement of the steering capability of the mechanism. The improved platform is shown in Figure 8.1. The abdominal rudder plate is added to the rear end below the battery compartment, and by changing the offset angle, asymmetric hydrodynamic moments on

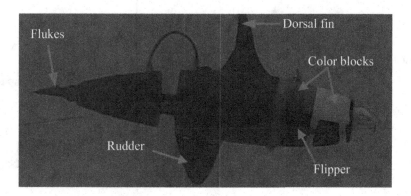

FIGURE 8.1 Prototype overview of the gliding underwater robot.

both sides can be generated. In addition, the top of the pectoral fin cabin is affixed with color-coded blocks of different colors, which are mainly used for the global vision system to measure its plane motion state and record data.

For the robotic dolphin, the flippers are the only turning mechanisms that can provide the yaw moment. However, the thrust generated by flippers is relatively deficient, which directly leads to poor anti-interference ability. In practical applications, it is apt for the gliding underwater robot to apply the median and/or paired fin (MPF)–based turning modes in a narrow area. Conversely, the robot can achieve high speed under body and/or caudal fin (BCF) motion. By virtue of the single flipper's deflection, the robot can steer in BCF motion. Nevertheless, the turning performances are unsatisfactory just via the flippers, such as low turning rate and large turning radius, the reason for which may lie in two aspects. On one hand, the flipper's water supply area is not big enough to generate a large hydrodynamic force. On the other hand, the flippers locate at a position closer to the center of gravity and are not along the axis of the body. Thereby, a rudder, owing to its similar shape to flippers, is newly designed to install on the belly of the gliding underwater robot to improve the turning ability. Combining the rudder with flipper, we propose five turning modes, including the two typical BCF-based and three typical MPF-based turning patterns, as illustrated in Figure 8.2. The two-way arrow indicates

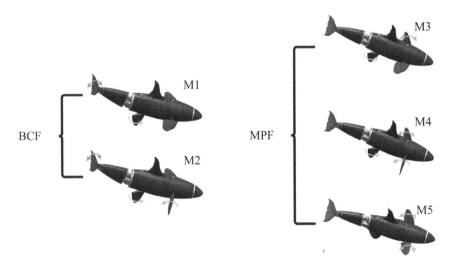

FIGURE 8.2   The schematic diagram of five turning motions.

that the joint can continuously flap, while the one-way arrow illustrates that the joint can only deflect to a fixed angle. The detailed introductions of five modes are taken as follows:

- *BCF-based mode*: The first mode, labeled M1, is just on the basis of the rudder to obtain the yaw moment. Meanwhile, the waist and caudal joints are driven by central pattern generators (CPG) model. When the rudder is deflected to a negative angle, the gliding underwater robot will acquire the yaw moment to turn right. Furthermore, the flipper joins in the turning task in the second mode, M2. Obviously, if the right flipper is deflected to −90°, the robot can get a larger yaw moment. Hence, by combining the flipper with the rudder, the turning ability can be significantly improved.

- *MPF-based mode*: The third and fourth modes are nearly the same as the first and second ones, other than that the BCF mode is replaced by the MPF mode of the unilateral flipper. As for the fifth mode M5, the robot flaps two flippers simultaneously, in which the right flipper is deflected to 180° first. Via differential flapping of bilateral flippers, the two flippers can generate reverse direction hydrodynamic forces, and thereby, the robot can steer with quite a little radius.

In theory, the turning radius with BCF-based turning mode may be larger than that with MPF-based mode, the main reason for which is that the forward speed of BCF mode is significantly greater. In particular, it is worthwhile to mention that we just conduct the experiments in which the robot turns around clockwise due to the limitation of a small test pool. Regarding the counterclockwise turning motion, the conditions are the same owing to the symmetry of the mechanical mechanism.

8.2.2 Vertical Motion

High maneuverability in the vertical plane is of crucial importance for the AUVs to realize efficient 3-D motion. Thanks to the powerful pitch adjustment capability of the propulsive mechanism, the robotic dolphin has a natural advantage in vertical motion. Furthermore, with the support of the gliding parts, the gliding underwater robot owns a stronger ability for accomplishing trickier tasks in real-world applications through a combination of dolphin-like and gliding motions. For traditional AUVs, the ways of vertical motion are mostly subject to insufficient maneuverability.

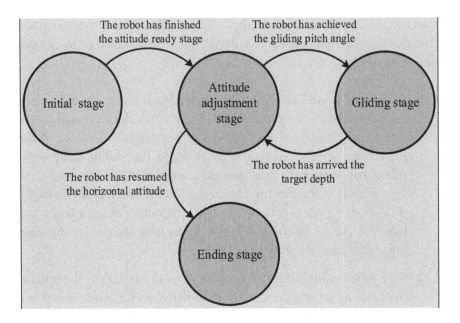

FIGURE 8.3   The finite state machine of the vertical motion.

Therefore, we propose a finite state machine (FSM) framework to realize the gliding motion with a large gliding angle, as illustrated in Figure 8.3. There are four stages in the FSM framework.

- *Initial stage*: In this stage, the robot does some preparatory works including the initialization of buoyancy-driven, centroid, and pro-pulsive systems. When the pitch angle exceeds a threshold, the system will enter the next stage.

- *Attitude adjustment stage*: The task of this stage is to change the attitude rapidly with the help of the propulsive system. Hence, there are two conditions at this stage. On one hand, when the initialization is finished, the BCF mode is started up with CPG offset to adjust the pitch angle. Note that we need to add the offset β_i to output angle since we choose the unilateral oscillators as the CPG model output; that is, $y_i = \psi_i + \beta_i$. ψ_i indicates the output angle calculated by CPG. On the other hand, when the robot glides to a target depth or has to glide up, the robot will adjust to horizontal or vertical upward atti-tude, and the stage will be executed.

- *Gliding stage*: When obtaining a larger pitch angle, the robot can switch to the gliding stage. In this stage, the robot can exert the closed-loop control to keep a larger gliding angle.

- *Ending stage*: When the robot has adjusted to the target attitude after the second stage, the vertical motion has been completed. The next task can be carried out further.

8.2.3 Experimental Results and Analysis of Horizontal Motion

To assess the 3-D motion performances of the gliding underwater robot, extensive experiments are conducted in a 1.1-m-depth test pool. In particular, through identifying the color block on the gliding underwater robot, a global camera system is employed to measure its motion.

In this subsection, we design four sets including the rudder angle δ_u, flipper angle δ_f, and flapping frequency of two modes f_b, f_m, which takes the forms as

$$
\begin{cases}
\Delta_U = \{0°, 30°, 45°, 60°\} \\
\Delta_F = \{0°, 90°\} \\
F_B = \{1\,\text{Hz}, 1.5\,\text{Hz}, 2\,\text{Hz}\} \\
F_M = \{1.5\,\text{Hz}, 2\,\text{Hz}, 2.5\,\text{Hz}\}
\end{cases}
\tag{8.1}
$$

8.2.3.1 BCF-Based Turning Mode

The experimental results are plotted in Figure 8.4, in which Figure 8.4a represents the data under different δ_u and δ_f. The data on the left side of Figure 8.4a show the robot's trajectory, from which we can draw some conclusions. First, when combining the rudder with flipper, the turning radii were much smaller than using either one. Second, the turning performance was better when $\delta_u = 45°$, the main reasons for which may lie in larger water supply area than in the condition of $\delta_u = 30°$, and relatively higher speed than in $\delta_u = 60°$. As for the trajectory when $f_b = 1.5\,\text{Hz}$, $\delta_u = 0°, \delta_f = 90°$, it seemed that the robot occurred deflection earlier, the reason for which was the non-zero initial yaw angle. As expected, it owned a lower turning rate in the latter stage. Regarding the yaw angle, we could conclude the order of average turning rate.

$$
\begin{aligned}
W_{\delta_u=45°,\delta_f=90°} &> W_{\delta_u=60°,\delta_f=90°} > W_{\delta_u=45°,\delta_f=0°} \\
&> W_{\delta_u=30°,\delta_f=90°} > W_{\delta_u=30°,\delta_f=0°} > W_{\delta_u=60°,\delta_f=0°}
\end{aligned}
\tag{8.2}
$$

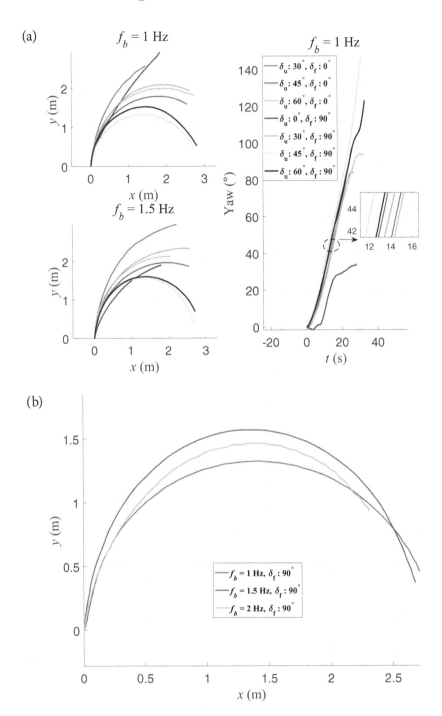

FIGURE 8.4 Experimental data of turning mode based on BCF. (a) Under different rudder angles. (b) Under different CPG frequencies.

where $\delta_u = 0°, \delta_f = 90°$, the turning performance was poor. Of course, the factors influencing the turning rate were manifold, such as the forward speed and swimming attitude. The general order could provide a theoretical basis for further deep research.

Figure 8.4b denotes the results under different f_b when $\delta_u = 45°$. From the perspective of turning rate, after the robot enters the steady state, the conclusion that the robot had better performances with higher frequency when the $\delta_f = 90°$ could be drawn.

8.2.3.2 MPF-Based Turning Mode

There are two modes based on MPF, the results of which are illustrated in Figure 8.5. The performance conditions of unilateral MPF-based mode are similar to the ones of BCF-based mode, other than smaller turning radii less than 1 body length (BL). Interestingly, the robot could steer with a little turning radius as small as approximately 0.2 BL under the bilateral

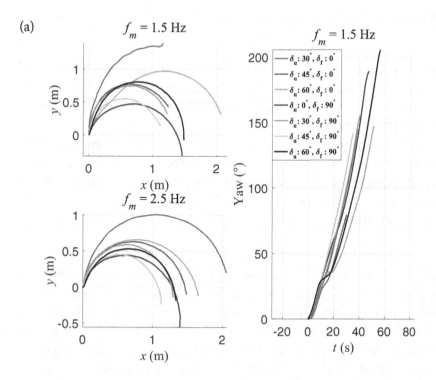

FIGURE 8.5 Experimental data of turning mode based on MPF. (a) Flapping of unilateral flipper. (b) Flapping of bilateral flippers.

(b)

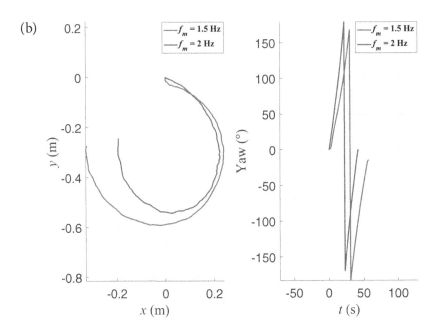

FIGURE 8.5 (*Continued*)

MPF-based mode. Besides, as flapping frequency increases, the turning radius and turning rate were also optimized. Therefore, it is more suitable to apply the MPF-based mode in a narrow area where underwater operations often occur.

Through the preceding analysis of various turning modes, the performances of the gliding underwater robot in horizontal motion are investigated. Every turning pattern has its own pros and cons. By combining these patterns, the robot can select different modes to achieve steering in various situations.

8.2.4 Experimental Results and Analysis of Vertical Motion

According to the FSM, we carried out some simulations and experiments to verify the effectiveness of the framework. Figure 8.6 indicates the simulation results in which the gliding underwater robot accomplishes a complete gliding round. In particular, compared with the common gliding motion, a much larger pitch angle, more than $80°$, was reached when the robot entered the steady gliding state. From the point of gliding angle, the gliding underwater robot could realize approximately $65°$, which validated the effectiveness of the proposed framework. In particular, before

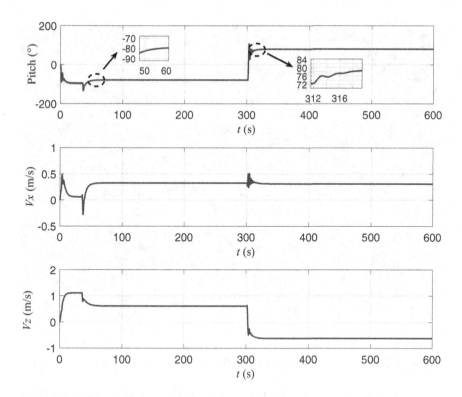

FIGURE 8.6   Simulation results of vertical motion.

$t = 35$ s, the robot dived under a dolphin-like motion and basically realized the vertical downward motion. Therefore, when there are tasks that require the robot to dive with a vertical downward attitude, the dolphin-like motion can be selected. Actually, the attitude of robot in FSM-based gliding motion was also closer to the vertical downward.

Due to the size limitation of the test pool, we conducted some experiments of dolphin-like motion to rapidly adjust the attitude, which corresponded to the first and second stages in the FSM framework. The snapshot sequences are figured in Figure 8.7. In particular, the ith figure of Figure 8.7 denotes the trajectory of vertical motion. As we can see, the gliding underwater robot took less than 10 s to complete the action. Besides, we also analyze the attitude information from the sensor, which indicates that the roll angle was basically the same in the whole process. Moreover, from the point of pitch angle of attitude sensor, there was a large acceleration since we set the waist and caudal angles to the opposite direction in the beginning, which resulted in a large pitch moment when starting the CPG flapping.

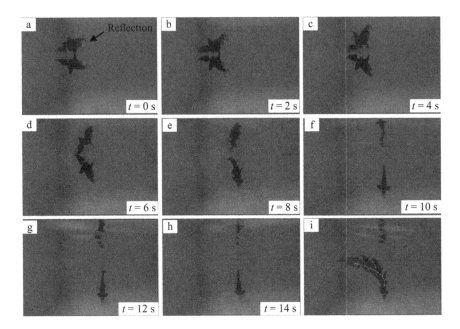

FIGURE 8.7   Snapshot sequences of vertical motion. (a) $t = 0$ s, (b) $t = 2$ s, (c) $t = 4$ s, (d) $t = 6$ s, (e) $t = 8$ s, (f) $t = 10$ s, (g) $t = 12$ s, (h) $t = 14$ s, (i) trajectory of vertical motion.

Considering the gliding underwater robot has a shortage of capabilities in yaw motion, the performance in horizontal motion has been greatly improved with the aid of the rudder and the flipper. By virtue of the combination of five patterns, the gliding underwater robot can accomplish tasks better according to different scenarios. Note that there are two more BCF-based modes, which promote the practical applications of the gliding underwater robot. Furthermore, an FSM framework is offered to achieve the vertical downward action and large gliding angle, which enhances the ability of vertical motion. However, for horizontal motion, the questions of how to combine and switch these patterns are of great need for better maneuverability. Similarly, in respect of the switching between the dolphin-like and gliding motions in the vertical plane, we simply reset the propulsive system as the transition way in this work, which may cause an unstable attitude as shown in Figure 8.6. Therefore, it is worth exploring the transition procedure between two motions.

8.3 3-D PATH PLANNING WITH MULTIPLE MOTIONS

The 3-D maneuverability analysis in the previous section shows the motion performance of the gliding underwater robot in the yaw and pitch directions. It not only obtains the maneuverability relationship and turning

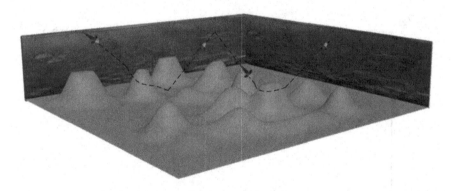

FIGURE 8.8   Schematic diagram of 3-D path planning in an ocean environment.

radius of various yaw motions but also verifies that the gliding underwater robot is in the vertical direction. The ability to perform vertical motion in a large pitch attitude. Therefore, based on the previous experimental conclusions, this section presents a multimodal 3-D path-planning framework suitable for the marine environment. Figure 8.8 shows a schematic diagram of 3-D path planning.

Considering that the underwater robot needs to meet the requirements of long-range and high maneuverability at the same time when it operates in the vast sea, the proposed framework makes full use of the multimodal characteristics of the gliding underwater robot and divides the 3-D path planning into the cruise phase and the obstacle avoidance phase. The former uses the gliding motion for path optimization, and the latter applies the dolphin-like motion to complete high maneuverability obstacle avoidance.

8.3.1 Gliding Path Planning Based on Geometric Constraint

For underwater gliders, the sawtooth-like path is a typical trajectory, as shown in Figure 8.9, which is determined by the gliding angle and diving depth. However, the gliding angle and diving depth are often constrained. On one hand, an excessive gliding angle may cause the robot to stall. On the other hand, due to the increasing of water pressure, deep diving depths may lead to material deformation and mechanical structure damage. Therefore, we set the constraints of the gliding angle, diving depth, and turning radius.

$$\begin{cases} D \leq D_{max} \\ \gamma \leq \gamma_{max} \\ R \geq R_{min} \end{cases}, \qquad (8.3)$$

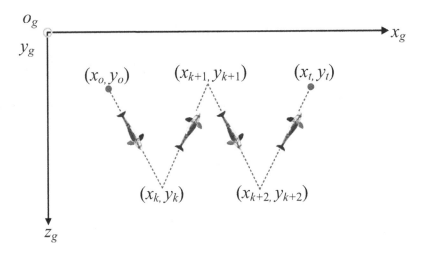

FIGURE 8.9   Sawtooth-like path.

where D_{max} and γ_{max} represent the maximum depth and gliding angle of gliding underwater robot, respectively; R_{min} denotes the minimum turning radius constraint.

Furthermore, the starting and ending points of a sawtooth-like path are set to $Point_o = (x_o, y_o)$ and $Point_t = (x_t, y_t)$ which are on the water surface. L refers to the horizontal distance between two adjacent way points. d represents the horizontal distance between the starting and ending points. n denotes the gliding round that is a positive integer. Thus,

$$L = \frac{D}{\tan \gamma} \text{ and } d = \|Point_t - Point_o\| = 2nL. \tag{8.4}$$

Due to the constraints of gliding angle and diving depth, we can derive a conclusion about n.

$$\begin{cases} n \le \dfrac{d \tan \gamma_{max}}{2D_{set}} & (if\ D = D_{set}) \\[2mm] n \ge \dfrac{d \tan \gamma_{set}}{2D_{max}} & (if\ \gamma = \gamma_{set}) \end{cases}$$

The gliding angle in the steady state is related to the lift–drag ratio. The greater the lift–drag ratio, the smaller the absolute value of the gliding angle. Furthermore, it can cause a smaller slope for the gliding trajectory and a longer gliding distance. Therefore, to some extent, a smaller gliding angle can achieve better gliding efficiency. However, a smaller gliding angle may lead to the shallow depth for our specific task. In this chapter, we set the maximum gliding angle at 43°, which is a comparative parameter in terms of gliding distance and diving depth. Moreover, maximum diving depth is set to 300 m. Hence, in accordance with the principle of the minimum number of glide rounds, the gliding round can be derived.

8.3.2 Obstacle-Avoidance Path Planning Based on Dolphin-Like Motion

As a key indicator in 3-D path planning, effective obstacle-avoidance planning is important to guarantee the safe navigation in complex seabed terrain. Next, the obstacle avoidance path planning of the gliding underwater robot will be introduced.

8.3.2.1 Characteristics Analysis of Obstacle Avoidance

The multimodal characteristics of the gliding underwater robot provide a variety of ideas for obstacle avoidance path planning, but the motion characteristics of the gliding and dolphin-like motion are different. Therefore, this section combines simulation data to analyze and discuss obstacle avoidance in different motion performances, providing data support for obstacle-avoidance path-planning research. In the field of marine operations, sonar is the most commonly used to detect environmental information, which can obtain the orientation and distance of the robot and the detected object. Using this feature, we puff the detected obstacles into spheres and set a safe threshold distance. Since the obstacles in this section are mainly on the bottom of the sea, the yaw direction is selected for the obstacle avoidance movement. Figure 8.10 shows the plane schematic diagram of the designed obstacle avoidance path. In order to improve the accessibility of the planned obstacle avoidance path, the minimum turning radius of the gliding underwater robot needs to be considered to ensure that the curvature radius of the planned path satisfies this constraint. Therefore, for the steering task shown in Figure 8.10, this section simulates the gliding and dolphin-like motions based on the complete dynamic model to provide a constraint reference for obstacle-avoidance path planning. It is worth noting that, according to the 3-D maneuverability analysis, when only the unilateral flipper offset is retained, the steering radius of the gliding underwater robot is the largest. Therefore, this steering mode is applied in the simulation to ensure that the measured steering radius obtains maximum value in all motions.

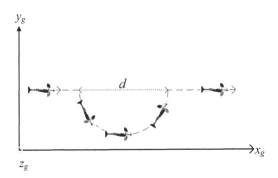

FIGURE 8.10   Illustration of motion task.

In gliding motion, the gliding underwater robot may not achieve the turning of large-scale angle with high speed due to poor maneuverability. When the robot enters a steady gliding state, the attitude of the fuselage will drastically deflect toward one direction if we exert the yaw moments by the flippers. Furthermore, the robot will gradually enter another equilibrium. However, on account of the small forward speed that may not provide enough hydrodynamic forces, it is difficult to adjust the flipper angles to make the robot deflect toward the opposite direction. Therefore, the switching process between turning states needs to be smooth and subtle, rather than turning at a large angle; otherwise, it may fall into a spiraling movement equilibrium that cannot be broken easily.

As shown in Figure 8.10, in order to accomplish this task, we need to set the deflection angle of the fuselage to $0° \sim 90° \sim 0° \sim -90° \sim 0°$. Therefore, we set up a step-by-step incremental approach to change the yaw angle and apply a simple PID controller:

$$\begin{aligned} \varphi(k) &= \varphi(k-1) + \Delta\varphi \\ \Delta\varphi &= \varphi_0 \cos\left((n+1)\pi\right) \end{aligned}, \tag{8.5}$$

where $n = 1, 2, 3, 4$ represent the four states, respectively. φ_0 is an artificial value that determines the growth rate. Finally, via the method, the target yaw curve and the tracking curve fit well.

Moreover, to explore the effect of the velocity $\|V_b\|$, we test the turning radius under different $\|V_b\|$ by adjusting the position of piston, further changing the net buoyancy. The simulation results for the projection of 3-D trajectory are shown in Figure 8.11. In Figure 8.11a, the turning radius increases as $\|V_b\|$ increases, and the movement in first stage becomes

(a)

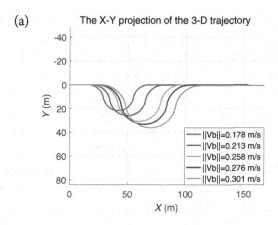

(b)

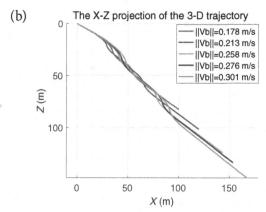

FIGURE 8.11 Projection of 3-D trajectory in gliding motion. (a) *X–Y*. (b) *X–Z*.

slower since the robot needs more time to generate enough yaw moments. Furthermore, within the velocity set of this chapter, it can be clearly found that the robot achieves a minimum turning radius, approximately 52 m, at a minimum velocity $\|V_b\|$, and the maximum radius is close to 100 m. Moreover, from Figure 8.11b, we can deduce that the gliding angle has always been around $40°$ and almost remains the same while turning, which provides the basis for the following obstacle avoidance tasks.

Regarding the dolphin-like motion, the gliding underwater robot can maintain a high speed with a small turning radius on account of its high maneuverability. It is also for this reason that the robot can complete the tasks faster. Unlike the gliding motion, the robot has a greater forward speed in dolphin-like motion and can adjust the body attitude more quickly. Therefore, in order to obtain a larger yaw moment, we set the deflection angle of active flipper to $90°$ during turning, rather than applying a feedback controller. For example, when the robot turns left, the deflection angles of left and right flippers are set to $90°$ and $0°$, respectively.

In the same way, we also conduct some simulations at different $\left\|V_b\right\|$, as shown in Figure 8.12. In particular, the velocity is achieved by changing the flapping frequency of flukes. Evidently, the velocity in the dolphin-like motion is much higher than that in the gliding motion. In addition, we can conclude that the minimum turning radius is 18 m when $\left\|V_b\right\| = 0.460 \, \text{m/s}$ from Figure 8.12a. It should be noticed that the maximum radius is 32 m when $\left\|V_b\right\| = 0.785 \, \text{m/s}$, which is smaller than the minimum one of gliding motion. With regard to the X-Z projection, we can regulate the pitch

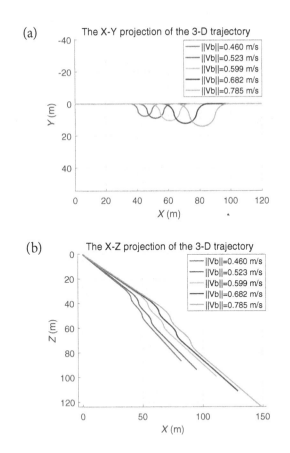

FIGURE 8.12 Projection of 3-D trajectory in dolphin-like motion. (a) X–Y. (b) X–Z.

angle by the movable slider. Therefore, the robot can achieve the same motion direction as the gliding motion, as shown in Figure 8.12b.

Through the above analysis, we can see that the gliding underwater robot can achieve the specific task of obstacle avoidance and is more suitable to complete tasks by dolphin-like motion. However, if the detection distance is large enough, the gliding motion can also be used to reduce energy consumption in a simple obstacle environment. Furthermore, through motion characteristics analysis of the full-state dynamic model, we also obtain the minimum turning radius and gliding angles data under two motions, which directly offer the motion constraints for 3-D path planning.

8.3.2.2 Local Obstacle-Avoidance Path Planning Based on A*

When the gliding underwater robot moves along the gliding path, it may encounter obstacles that can be detected by the sonar that owns a $R_s \leq 40$ m detection range. At this time, it is necessary to replan the path for avoiding obstacles and return to the original gliding path. Since the sonar can get the information about approximate shape of the obstacle and the distance to the obstacle, we simplify the obstacle into a sphere with maximum length of the detected obstacle as the initial diameter. In addition, in order to ensure that the obstacles can be avoided, a safety threshold d_0 is added to the initial diameter. Moreover, the d_0 will also be applied in path smoothing.

Once the obstacle is detected, the obstacle avoidance path should be planned in real time. First, we can obtain the initial point $Point_i = (x_i, y_i, z_i)$ and ending point $Point_e = (x_e, y_e, z_e)$ of the obstacle avoidance path by calculating the intersections of gliding path and extended spherical obstacle:

$$\begin{cases} \dfrac{x-x_0}{x_k-x_0} = \dfrac{y-y_0}{y_k-y_0} = \dfrac{z-z_0}{z_k-z_0} \\ (x-x_s)^2 + (y-y_s)^2 + (z-z_s)^2 = R^2 \end{cases}, \tag{8.6}$$

where (x_s, y_s, z_s) refers to the center of the sphere. For convenience and brevity, we define $Point_i = (Point_{ix}, Point_{iy}, Point_{iz})$, and the same explanation can be also applied to $Point_e$. In particular, due to the limitation of minimum turning radius, R can be obtained by

$$R = \max\{R_s, R_{\min}\}. \tag{8.7}$$

Here, we denote $d = d_0 + r$, where r denotes the radius of the initial sphere, and R_{min} can be obtained by motion analysis.

Furthermore, 3-D obstacle avoidance path needs to be planned. In recent years, the popular algorithms for path planning mainly conclude the following categories: potential field algorithms (PFAs), evolutionary algorithms (EAs), RRTs, graph search algorithms, and others. The A* algorithm is one of graph search algorithms, which improves the logic of graph search with heuristic evaluations. Among these algorithms, the PFA has an obvious tendency toward local minima if the environment is complex; therefore, it needs to be complemented by other methods [249]. The RRT has a shorter planning time than A*, but the generated path by RRT is fairly unsmooth [246] and usually not optimal. Hence, we select the A* algorithm as the planning method. The A* algorithm is a heuristic search algorithm based on Dijkstra [249][250]. It is an optimal first-priority search algorithm. In particular, it is effective for the optimal search of the road network. The key of A* is to determine the heuristic function, which takes the forms as follows:

$$f(n) = h(n) + g(n), \tag{8.8}$$

where $g(n)$ represents the cost from the starting point to the candidate node n. $h(n)$ indicates the cost from the candidate node n to the target point. Therefore, while ensuring the optimal path, it can also reduce the search complexity to some extent. Regarding the selection of $g(n)$, Manhattan distance, diagonal distance, and Euclidean distance are usually used. More important, the heuristic function $h(n)$ is the estimation of the cost of the optimal path from the candidate node to the target. Generally, in 2-D path planning, there are some well-known heuristic functions to use on a square grid map. The Manhattan distance is a standard heuristic function to employ in a grip map. Moreover, the Chebyshev, diagonal distance, and Euclidean distance are also often selected in practical applications. In addition, the Manhattan distance has higher calculation efficiency, which is a quite important factor for the gliding underwater robot to realize path planning on the embedded platform. In this chapter, by comparison with the three distances, the results show the planning via Manhattan distance is better than that via diagonal distance in aspect of efficiency, and the Manhattan distance performs on par with diagonal distance in terms of path length. Hence, we select the Manhattan distance as

$$h(n) = \left| n_x - Goal_x \right| + \left| n_y - Goal_y \right|, \tag{8.9}$$

where (n_x, n_y) and $(Goal_x, Goal_y)$ denote the coordinates of candidate node and target node, respectively. However, A* algorithm is not suitable for 3-D path planning due to high searching complexity. Hence, in this chapter, we offer an improved A* algorithm which plans the 3-D path in the 2-D horizontal and vertical planes, respectively. First, we project the 3-D path onto a 2-D horizontal plane, as shown in Figure 8.13. Next, a new plane coordinate frame C_a is established, and the 2-D A* algorithm is employed to obtain the obstacle avoidance path. Furthermore, when mapping the 2-D waypoints on C_a back to C_g, it should be noted that a rotation matrix is required since the y-axis of the two coordinate systems is not parallel. Besides, translation matrix should also be calculated. Therefore, we can obtain the augmented vector of the 3-D coordinates $\widehat{Point}_{3D}$ according to the transformation. Moreover, we denote the first two elements of the vector as the $Point_{3D(x,y)}$.

$$\widehat{Point}_{3D} = Tr \cdot \widehat{Point}_{2D}, \tag{8.10}$$

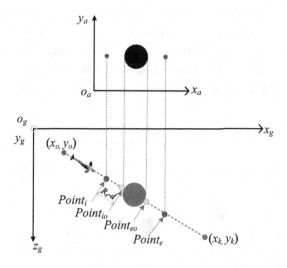

FIGURE 8.13   Illustration of path projection.

where

$$Tr = \begin{pmatrix} \cos\delta & -\sin\delta & 0 & Point_{ix} \\ \sin\delta & \cos\delta & 0 & Point_{iy} \\ 0 & 0 & 1 & 0 \\ 0 & 0 & 0 & 0 \end{pmatrix}.$$

$$\widehat{Point}_{2D} = \begin{pmatrix} x & y & 0 & 1 \end{pmatrix}^{T}$$

$$\delta = \tan^{-1}\left(\frac{Point_{ey} - Point_{iy}}{Point_{ex} - Point_{ix}}\right)$$

Since the path replanned by A* is composed of discrete points, the iteration steps K can be obtained. Regarding the k–time coordinate value $Point_{3D(z(k))}$, a uniform linear distribution is employed.

$$\begin{cases} Point_{3D(z(k))} = Point_{iz} + \sum_{n=1}^{k} S(n) \\ S(n) = \dfrac{Point_{ez} - Point_{iz}}{K} \end{cases} \quad (8.11)$$

Afterward, due to obstacles that may appear in different positions of the gliding path, a variety of obstacle avoidance forms should be generated. When the obstacle is closer to way points of the gliding path, the ending point $Point_e$ should be redistributed. Depending on the distance of the obstacle from waypoints, we set three obstacle forms, as shown in Figure 8.14. It should be noted that these forms are only discussed for

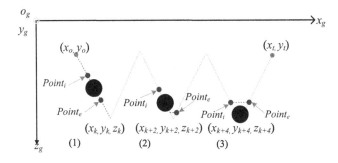

FIGURE 8.14   Illustration of three obstacle forms.

underwater way points in this chapter, and the situations of waypoints on the surface are nearly the same.

8.3.2.3 Path Smoothing Based on Bezier Curve

Since the path planned by A* is relatively twisty, it is necessary to smooth the path to meet the kinematics and dynamic constraints of the robot. Especially in the initial and ending stages of obstacle avoidance, if the path is not smooth, the robot needs to achieve large-angle steering in a short time, which may cause the mechanism to be damaged due to violent movements. The Bezier curve is usually utilized to smooth the twisted curve [251]. In this chapter, we propose a three-stage Bezier smoothing method, in which a segmented Bezier curve fitting A* path is implemented. Generally, a form of Bezier curve is defined as

$$B(t) = \sum_{i=0}^{n} P_i b_{i,n}(t), t \in [0,1], \tag{8.12}$$

where n and P_i are the degree and control points of Bezier curve, respectively. Furthermore, the ith basis function of order n, $b_{i,n}$, is defined as

$$b_{i,n}(t) = \binom{n}{i} t^i (1-t)^{n-i}, i = 0,...,n. \tag{8.13}$$

In this chapter, we divide the entire obstacle avoidance path into three segments. The first and third parts mainly deal with the transition, and the second part handles the path smoothing problem. In the first and third stages, we can see that the robot needs to achieve a $90°$ turning, which is beyond the robot's mobility. Therefore, we chose some control points to smooth the transition. In order to achieve continuous steering at a small angle in the initial and final stages of the obstacle avoidance path, one of the control points should be chosen at the intersection of the gliding path and the obstacle sphere. Regarding the second stage, the robot should achieve slow steering, and the control points are selected among the points of original path.

8.4 RESULTS AND ANALYSES

In order to evaluate the effectiveness of the proposed method, we conducted simulation experiments. The simulations were carried out in MATLAB®/Simulink using the full-state dynamic model.

8.4.1 Result of Gliding Path Generation

In the part of gliding path generation, we first need to set the starting and ending points that determine the gliding rounds, gliding angle, and gliding depth. Figure 8.15 represents the 3-D gliding paths under three cases. The starting point for the first case is set to $(10, 20, 0)^T$, then the other two cases are connected end to end. The ending points of the second and third cases are $(500, 1000, 0)^T$ and $(1500, 3000, 0)^T$, respectively. Regarding the gliding depth in three cases, they are 93.83 m, 174.70 m, and 291.17 m, respectively. On the other hand, the results of gliding angle are 43°, 38°, and 38°, respectively.

In the first case, due to the small distance between the starting and ending points, only one gliding round can be completed, which directly leads to a large gliding angle and a shallow gliding depth. In the other two cases, in accordance with the principle of minimum gliding round number and optimal gliding angle, the gliding depth can be calculated.

8.4.2 Result of Obstacle Avoidance

When the gliding underwater robot moves along the generated gliding path, it will detect the obstacles in real time. Once the obstacles are found, the obstacle avoidance path is calculated by the 3-D improved

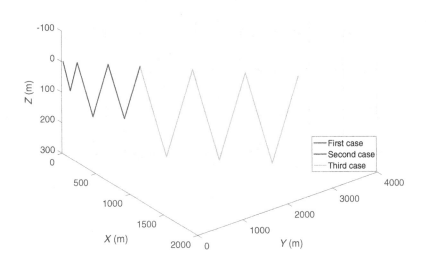

FIGURE 8.15   Gliding path under three cases.

A* algorithm. The earlier gliding parameters are employed again to test the obstacle avoidance. The difference is that the ending point is fixed at $(100, 200, 0)^T$ to facilitate better observation of obstacle avoidance results. Besides, we set the radius of the initial sphere $r = 20$ m, and the safety threshold $d = 5$ m.

Furthermore, in the part of real-time obstacle avoidance, we set three obstacle forms in accordance with the positions of obstacles in the gliding path. Figure 8.16 shows the results when the gliding robotic encounters obstacles in three forms, respectively.

In the case of the first type of obstacle, since the ending point of the obstacle-avoidance path does not cross the way points of gliding path, the robot moves along the path planned by A* and returns to the gliding path. In the second case, the $Point_e$ crosses the way point while $Point_{eo}$ does not. Thereby, the ending point of obstacle avoidance path is set at the central symmetry point of $Point_{eo}$ with respect to the way point. In the last case, when both $Point_e$ and $Point_{eo}$ cross the waypoint, we select the ending point at the central symmetry point of $Point_e$ with respect to the waypoint.

It can be seen that the path planned by A* is almost along the threshold ring. Besides, in the second and third cases, the ending point of the arrival is somewhat different from the setting point. The main reason is that the map grid needs to be divided in the A* algorithm. In order to ensure real-time performance, we divide the map grid into 50×50. Therefore, when the total number of points on the obstacle avoidance path is larger than the meshing range, a certain deviation will occur. In fact, the distance differences between the ending points in two cases are 2.1336 m and 1.1272 m, which account for 1% and 0.5% of the total horizontal distance L, respectively. In particular, it should be noted that Figure 8.16 presents the results when the obstacle appears in the diving part of the gliding path. Regarding the surface processing, the computational procedures are the same.

8.4.3 Result of Path Smoothing

As can be seen from the previous section, the resulting obstacle avoidance path is not smooth. For instance, the state of the yaw angle displays quite tortuously when the robot avoids obstacles without path smoothing, which causes the robot to turn frequently and damage the actuator.

Furthermore, the comparisons of attitude angles before and after smoothing in two cases are plotted in Figure 8.17, respectively. With regard to the third case, it is similar to the first case, other than that the pitch angle should keep constant while the robot avoids obstacles since $Point_i$

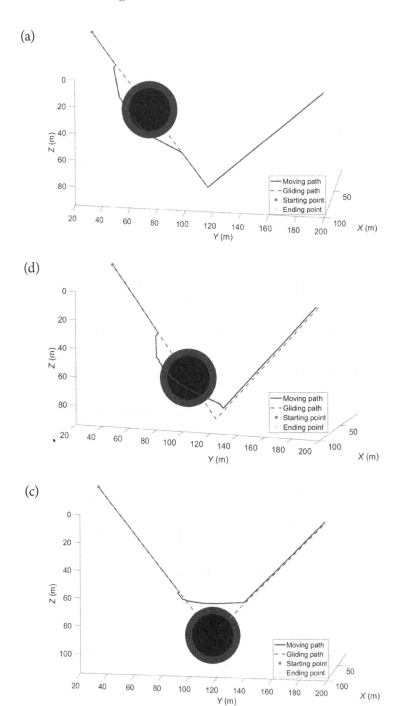

FIGURE 8.16   Obstacle avoidance illustration of 3-D trajectory. (a) First obstacle form. (b) Second obstacle form. (c) Third obstacle form.

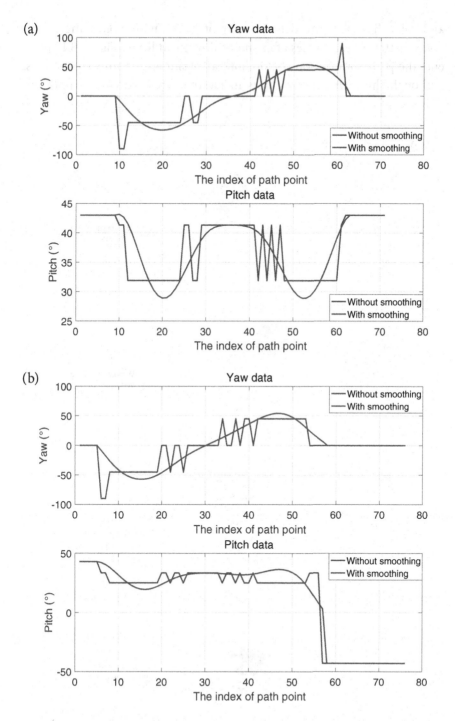

FIGURE 8.17 Change of yaw and pitch angle when path smoothing is performed.
(a) First case. (b) Second case.

and *Point_e* have the same depth. Apparently, the yaw and pitch angles represent better performances after smoothing. In addition, Figure 8.18 presents the path comparisons before and after smoothing under two cases. Although the path slightly enters inside the threshold ring, it does not

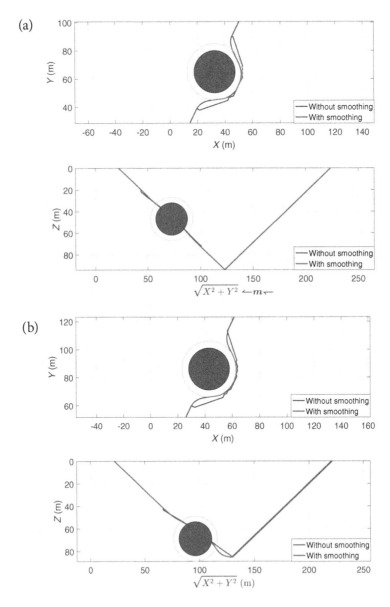

FIGURE 8.18 Path comparison before and after smoothing. (a) First case. (b) Second case.

TABLE 8.1   Model Parameters of the Depth Control System.

Case no.	Unsmooth length	Smooth length
Case 1	290.04 m	287.79 m
Case 2	278.63 m	275.88 m
Case 3	271.06 m	267.40 m

touch the obstacle ring. In particular, the projection curve of the Z-axis with smoothing in Figure 8.18b shows a flat transition around the floating switch point, which is more conducive to the attitude adjustment when the robot prepares to float. In terms of the total length of the path, the path after smoothing is slightly shorter than the path without smoothing, as tabulated in Table 8.1.

8.4.4 Discussion

A novel 3-D path-planning method is employed to avoid obstacles exerted on the gliding underwater robot and offers a new thought for the gliding underwater robot to execute real-time tasks in the complex ocean environment. Compared with the method used by Wu *et al.* [247], the path we designed is smoother and more practical. Even though two motions including gliding and dolphin-like motions need to be switched during the moving, the simulation results have proved their feasibility. In addition, it is due to the combination of two motions that the effectiveness and smoothness of the obstacle avoidance path are guaranteed. Furthermore, the elapsed time of the method in MATLAB is less than 100 ms while the path is planned under smoothing. For instance, the average elapsed time in Figures 8.18a and 8.18b are 71 ms and 65 ms, respectively.

Although the elapsed time can meet the computational real-timeliness requirement, the movement path after obstacle avoidance cannot completely match the original gliding path, as illustrated in Figures 8.16b and 8.16c. The reason for the phenomenon is that the map size $n = 50$ of the A* algorithm is a bit small. However, via setting the map size $n = 100$ and $n = 150$, we obtain the average elapsed time in the first case as 220 ms and 481 ms which cannot meet the real-time request to a certain extent. To address this issue, on one hand, we can try another faster path planning method, such as RRT, to replace the A* algorithm. On the other hand, the path smoothing method can be improved to save time. In addition, Table 8.1 reveals that the total length is not largely shorten after smoothing. By introducing the intelligent optimization algorithm and taking the path length as one of the optimization goals, this problem can fully be solved.

8.5 CONCLUDING REMARKS

In this chapter, the aim was the 3-D maneuverability and path planning of the gliding underwater robot suitable for the marine environment. First, the steering mechanism of the gliding underwater robot is improved. A multimodal yaw maneuvering strategy is designed to enrich its steering modes, and its yaw maneuver performance is verified through experiments. Second, a pitch maneuvering strategy based on s finite state machine is designed. The underwater experiment results show that the proposed method can make the gliding underwater robot achieve high vertical motion. Finally, based on the kinematic characteristics obtained from the 3-D maneuverability analysis, a multimodal 3-D path-planning framework is proposed, which mainly includes the global path planning of the gliding motion, obstacle avoidance based on the A* algorithm with dolphin-like motion, and path smoothing algorithm based on Bezier curve, and the effectiveness of the proposed method is verified by simulations.

REFERENCES

[243] R. Cui, Y. Li, and W. Yan, "Mutual information-based multi-AUV path planning for scalar field sampling using multidimensional RRT*," *IEEE Trans. Syst. Man Cybern. Syst.*, vol. 46, no. 7, pp. 993–1004, Jul. 2016.

[244] Z. Peng, J. Wang, Y. Li, and W. Ya, "Output-feedback path-following control of autonomous underwater vehicles based on an extended state observer and projection neural networks," *IEEE Trans. Syst. Man Cybern. Syst.*, vol. 48, no. 4, pp. 535–544, Apr. 2018.

[245] J. D. Hernandez, G. Vallicrosa, E. Vidalric, E. Pairet, M. Carreras, and P. Ridao, "On-line 3-D path planning for close-proximity surveying with AUVs," in *Proc. Int. Fed. Autom. Control, Girona*, Spain, Apr. 2015, pp. 50–55.

[246] K. Yang and S. Sukkarieh, "3-D smooth path planning for a UAV in cluttered natural environments," in *Proc. IEEE/RSJ Int. Conf. Intell. Robot. Syst.*, Nice, France, Sep. 2008, pp. 794–800.

[247] Z. Wu, M. Zhao, Y. Wang, Y. Liu, H. Zhang, S. Wang, and E. Qi, "Path planning for underwater gliders with motion constraints," in *Proc. Asian Conf. Mech. Mach. Sci.*, Guangzhou, China, Dec. 2016, pp. 3–10.

[248] J. Cao, J. Cao, Z. Zeng, B. Yao, and L. Lian, "Toward optimal rendezvous of multiple underwater gliders: 3-D path planning with combined sawtooth and spiral motion," *J. Intell. Robot. Syst.*, vol. 85, no. 1, pp. 1898–206, Jan. 2017.

[249] J. Tan, L. Zhao, Y. Wang, Y. Zhang, and L. Li, "The 3-D path planning based on A* algorithm and artificial potential field for the rotary-wing flying robot," in *Proc. Int. Conf. Intell. Human Mach. Syst. Cybern.*, Hangzhou, China, Sep. 2016, pp. 551–556.

[250] H. Pan, C. Guo, and Z. Wang, "Research for path planning based on improved astart algorithm," in *Proc. Int. Conf. Inf. Cybern. Comput. Soc. Syst.*, Dalian, China, Jul. 2017, pp. 225–230.

[251] C.-C. Lin, W.-J. Chuang, and Y.-D. Liao, "Path planning based on Bezier curve for robot swarms," in *Proc. Int. Conf. Genet. Evol. Comput.*, Kitakyushu, Japan, Aug. 2012, pp. 253–256.

Printed in the United States
by Baker & Taylor Publisher Services